Preparation and Properties of 2D Materials

Preparation and Properties of 2D Materials

Special Issue Editors

Byungjin Cho
Yonghun Kim

MDPI • Basel • Beijing • Wuhan • Barcelona • Belgrade • Manchester • Tokyo • Cluj • Tianjin

Special Issue Editors

Byungjin Cho
Chungbuk Naiotnal University
Korea

Yonghun Kim
Korea Institute of Materials Science
Korea

Editorial Office
MDPI
St. Alban-Anlage 66
4052 Basel, Switzerland

This is a reprint of articles from the Special Issue published online in the open access journal *Nanomaterials* (ISSN 2079-4991) (available at: https://www.mdpi.com/journal/nanomaterials/special_issues/2D_nano).

For citation purposes, cite each article independently as indicated on the article page online and as indicated below:

LastName, A.A.; LastName, B.B.; LastName, C.C. Article Title. *Journal Name* **Year**, *Article Number*, Page Range.

ISBN 978-3-03936-258-5 (Pbk)
ISBN 978-3-03936-259-2 (PDF)

Contents

About the Special Issue Editors

Byungjin Cho received his MSc degree and PhD in materials science and engineering from the Gwangju Institute of Science and Technology in 2007 and 2012. He worked as a postdoctoral researcher at UCLA, USA, in 2012. After this, he was a senior researcher at Korea Institute of Materials Science (KIMS), Korea, from 2013 to 2016. Since 2017, he has been working as an assistant professor at Chungbuk National University. He has published dozens of SCIE papers on the synthesis of emerging 2D nanomaterials and nanoelectronics, such as sensor, transistor, and synapse devices.

Yonghun Kim received his MSc degree and PhD in materials science and engineering from the Gwangju Institute of Science and Technology in 2011 and 2016, and he was a postdoctoral researcher at the Korea Institute of Materials Science, Korea, in 2017. He has been a senior researcher at the Korea Institute of Materials Science since 2017. He has published dozens of papers on the synthesis of emerging two-dimensional nanomaterials, semiconductor integrated processes, neuromorphic synapse devices and advanced electrical characterization and reliability analysis.

Editorial

Preparation and Properties of 2D Materials

Byungjin Cho [1,*] and **Yonghun Kim** [2,*]

[1] Department of Advanced Material Engineering, Chungbuk National University, Chungdae-ro 1, Seowon-Gu, Cheongju, Chungbuk 28644, Korea

[2] Materials Center for Energy Convergence, Surface Technology Division, Korea Institute of Materials Science (KIMS), 797 Changwondaero, Sungsan-gu, Changwon, Gyeongnam 51508, Korea

* Correspondence: bjcho@chungbuk.ac.kr (B.C.); kyhun09@kims.re.kr (Y.K.); Tel.: +82-(0)43-261-2417 (B.C.); +82-(0)55-280-3281 (Y.K.)

Received: 19 March 2020; Accepted: 14 April 2020; Published: 16 April 2020

Since the great success of graphene, atomically thin layered nanomaterials—called two-dimensional (2D) materials—have attracted tremendous attention due to their extraordinary physical properties. In particular, van der Waals heterostructured architectures based on a few 2D materials, named atomic scale Lego, have been proposed as unprecedented platforms for the implementation of versatile devices with a completely novel function or extremely high performance, shifting the research paradigm in materials science and engineering [1]. Thus, diverse 2D materials beyond existing bulk materials have been widely studied for promising electronic, optoelectronic, mechanical, and thermoelectric applications. In particular, this Special Issue includes the recent advances in unique preparation methods, such as exfoliation-based synthesis and the vacuum-based deposition of diverse 2D materials, as well as their device applications based on their interesting physical properties. This editorial consists of the following two sections: Preparation Methods of 2D Materials and Properties of 2D Materials.

1. Preparation Methods of 2D Materials

Solution-based exfoliation methods for two-dimensional (2D) materials have been intensively investigated due to the ease of the process. In this regard, Zhang et al. investigated the cost-effective exfoliation method of multilayered 2D MoS_2 nanosheets and quantum dots from natural SiO_2-containing molybdenite in different solutions under mild ultrasonic conditions [2]. This simple method provides several advantages such as high yields, low cost and large-scale industrial perspectives compared with conventional methods. 2D-MoS_2 nanosheets with dimensions of 50–200 nm were prepared. Furthermore, the excellent photoconductivity of the nanosheets under visible light was demonstrated in various solution conditions. Meanwhile, the conventional method to prepare saturable absorber materials uses the Langmuir–Blodgett (LB) technique, the merits of which include its low cost. In this respect, Wang et al. demonstrated a low-cost reflective WS_2 saturable absorber (SA) on a silver-coated mirror for the first time [3]. By using the simple LB method, a large-area and highly uniform 2D-WS_2-coated SA was successfully shown. Moreover, the optical saturation properties of WS_2 SA were thoroughly analyzed, with the duration being around 409 ns and the highest peak power being 5.2 W. Thus, highly reflective WS_2 SA, created using the simple LB method, could be used in a diverse optical modulator with a wavelength of 1.3 μm.

MoO_3 is a promising material with well-recognized applications such as electronics, photocatalysis, electrocatalysis, batteries, and pseudocapacitors [4]. Among the various crystal structures of MoO_3, the orthorhombic α-MoO_3 provides unique 2D morphologies with layered structures. α-MoO_3 has been conventionally obtained via the hydrothermal method or sputtering. However, such conventional preparation methods have faced some critical challenges related to substantial energy, complex equipment, and expert operational skills. Thus, Ramos et al. report a new preparation method to obtain highly crystalline α-MoO$_3$ using vapor-phase synthesis [5]. They obtained highly ordered multilayer α-MoO$_3$ from molybdate using carbon nitride (g-C$_3$N$_4$) with a lamellar template. This simple method may be applied to electrocatalytic hydrogen evolution and ultrasensitive plasmonic biosensing.

For two-dimensional transition metal dichalcogenides (TMDCs), a uniform growth technique is required, especially for applications in electronics and optoelectronics. However, several critical challenges such as high growth temperature, limited growth area, and layer controllability still remain. Thus, Zhong et al. reported the simple growth method of 2D-MoS$_2$ using a two-step process, combining radio frequency (RF) magnetron sputtering and the subsequent sulfurization process [6]. The growth temperature of this two-step process is lower than 600 °C, and the crystalline qualities are simply controlled by RF sputtering power. As RF plasma power increases from 10 to 150 W, the crystalline quality also increases, which is confirmed by the intrinsic peak intensities of the Raman spectrum. Recently, a new family of 1D nanomaterials with weak van der Waals interactions was also reported. Kim et al. successfully demonstrated the synthesis of a 1D semiconductor V$_2$Se$_9$ crystal using mass production via the simple transport preparation method [7]. The 1D-V$_2$Se$_9$ crystal exhibited weak van der Waals interaction and a nanoribbon structure. Also, scanning Kelvin probe microscopy (SKPM) analysis showed a variation in work function depending on the thickness of the V$_2$Se$_9$ crystal. This mass-production preparation method for 1D nanomaterials such as V$_2$Se$_9$ could be suitably applied to the metal contact of future van der Waals-based nanoelectronic devices.

2. Properties of 2D Materials

The electrical properties of 2D semiconducting materials are usually validated via the demonstration of field effect transistor (FET) devices. Thus, research themes involving the performance enhancement of FET devices have long attracted great attention, especially with respect to device junction optimization. In this context, Lim et al. proposed a novel FET structure consisting of a 2D MoS$_2$/black phosphorous (BP) heterojunction, which shows a high on/off ratio of over 1×10^7, along with an extremely low subthreshold swing value of ~54 mV/dec and very low off current of ~fA level [8]. Interestingly, the low off current was attributed to the depletion region in the BP layer. Meanwhile, a TiO$_2$ interfacial layer inserted between a metal and 2D TMDCs (MoS$_2$ and WS$_2$) can also lead to enhanced FET properties [9]. In addition, a stable electrical performance could be achieved under a gate bias stress condition, since the TiO$_2$ interfacial layer serves as a Fermi level depinning layer, which reduces the density of the interface states.

The synthesis of p-type MoS$_2$ is often essential for the complementary integration process using p- and n-type 2D materials. Lee et al. reported that p-type semiconducting characteristics can be obtained via the addition of a dopant precursor of phosphorous pentoxide during the chemical vapor deposition synthesis process of MoS$_2$ [10]. The p-doped monolayer MoS$_2$ showed p-type conduction with a relatively low field effect mobility of 0.023 cm^2/V·s and an on/off current ratio of 10^3, compared with the pristine n-type MoS$_2$. The performance of the p-doped FET should be further improved. Along with neuron devices, artificial synapse devices have been recently considered as one of the most essential components in implementing a neuromorphic hardware system. Finding a physical parameter that precisely modulates synaptic plasticity is particularly required. Following this motivation, Kim et al. proposed a novel two-dimensional transistor architecture consisting of a NbSe$_2$/WSe$_2$/Nb$_2$O$_5$ heterostructure [11]. NbSe$_2$, WSe$_2$, and Nb$_2$O$_5$ function as a metal electrode, an active channel, and a conductance-modulating layer, respectively. Notably, the post-synaptic current was successfully modulated by the thickness of the interlayer Nb$_2$O$_5$, whose introduction facilitated the realization of reliable and controllable synaptic devices.

The unique optical properties of the 2D materials were intensively investigated. For instance, using thin semiconductor MoS$_2$/ferroelectric lead zirconate titanate heterostructure films, reversibly tunable photoluminescence was demonstrated during ferroelectric polarization reversal using nanoscale conductive atomic force microscopy tips [12]. The spontaneous polarization of the ferroelectric thin films affects the optoelectronic behaviors of MoS$_2$ indirectly via reversible electrochemical processes.

Meanwhile, the Raman spectrum of BP transferred onto a germanium-coated polydimethylsiloxane flexible substrate was systematically studied [13]. The Raman spectra obtained from several BP layers with different thicknesses showed the clear peak shifting rates for the Ag^1, B^2g, and Ag^2 modes. A study of the strain–Raman spectrum relationship was also conducted, showing a maximum uniaxial strain of 0.89%. The peak shifting of Ag^1, B_{2g}, and Ag^2 caused by this uniaxial strain was clearly measured. In another optical study, a systematic investigation of photoluminescence (PL) and Raman spectroscopy of the transferred bilayer-stacked MoS_2 were conducted, and compared with freestanding monolayer MoS_2 [14]. The interlayer difference and spatial inhomogeneity of exciton and phonon performance are attribute to film–substrate coupling-induced strain and doping. Even surface fluctuations with a thickness of less than one atom layer could be easily identified by Raman and PL spectroscopy, offering useful information about the 2D van de Waals homostructure and heterostructures' effects on the optical properties of 2D materials.

The mechanical properties of the 2D materials are also interesting and attractive. In this regard, the tribological performance of two kinds of WS_2 nanomaterials as additives in paraffin oil was investigated, showing that the friction and wear performance of paraffin oil can be greatly improved with the addition of WS_2 nanomaterials, and that the morphology and content of WS_2 nanomaterials have a significant effect on the tribological properties of paraffin oil [15]. For instance, paraffin oil with WS_2 nanoflowers exhibited better tribological properties than that with WS_2 nanoplates. The superior tribological properties of the WS_2 nanoflowers were attributed to their special morphology, which contributes to the formation of a uniform tribofilm during the sliding process.

We hope this Special Issue will help 2D material researchers follow up the latest research trends and progress in the 2D research community.

Author Contributions: B.C. and Y.K. wrote and revised the paper. All authors have read and agreed to the published version of the manuscript.

Funding: This work was supported by the National Research Foundation of Korea (NRF) grant funded by the Korean government (MSIT; Ministry of Science and ICT) (No. 2020R1A2C4001739) and Fundamental Research Program (No. PNK6990 and 6670) of the Korea Institute of Materials Science (KIMS). This research was also financially supported by the Ministry of Trade, Industry and Energy (MOTIE) and the Korea Institute for Advancement of Technology (KIAT) through the National Innovation Cluster R&D program (P0006704_Development of energy saving advanced parts).

Acknowledgments: The Guest Editors highly appreciate the effort of all authors for publishing their works in this Special Issue. We are also grateful to the editorial assistants who have made the publication process of the Special Issue very smooth and efficient.

Conflicts of Interest: The authors declare no conflict of interest.

References

1. Geim, A.K.; Grigorieva, I.V. Van der Waals heterostructures. *Nature* **2013**, *499*, 419–425. [CrossRef] [PubMed]
2. Zhao, W.; Jiang, T.; Shan, Y.; Ding, H.; Shi, J.; Chu, H.; Lu, A. Direct Exfoliation of Natural SiO_2-containing Molybdenite in Isopropanol: A Cost Efficient Solution for Large-scale Production of MoS_2 Nanosheetes. *Nanomaterials* **2018**, *8*, 843. [CrossRef] [PubMed]
3. Wang, T.; Wang, Y.; Wang, J.; Bai, J.; Li, G.; Lou, R.; Cheng, G. 1.34 μm Q-Switched $Nd:YVO_4$ Laser with a Reflective WS_2 Saturable Absorber. *Nanomaterials* **2019**, *9*, 1200. [CrossRef] [PubMed]
4. De Castro, I.A.; Datta, R.S.; Ou, J.Z.; Castellanos-Gomez, A.; Sriram, S.; Daeneke, T.; Kalantar-zadeh, K. Molybdenum Oxides—From Fundamentals to Functionality. *Adv. Mater.* **2017**, *29*, 1701619. [CrossRef] [PubMed]
5. Martín-Ramos, P.; Fernández-Coppel, I.A.; Avella, M.; Martín-Gil, J. α-MoO_3 crystals with a Multilayer Stack Structure Obtained by Annealing from a Lamellar MoS_2/g-C_3N_4 Nanohybrid. *Nanomaterials* **2018**, *8*, 559. [CrossRef] [PubMed]
6. Zhong, W.; Deng, S.; Wang, K.; Li, G.; Li, G.; Chen, R.; Kwok, H.-S. Feasible Route for a Large Area Few-Layer MoS_2 with Magnetron Sputtering. *Nanomaterials* **2018**, *8*, 590. [CrossRef] [PubMed]

7. Kim, B.J.; Jeong, B.J.; Oh, S.; Chae, S.; Choi, K.H.; Nasir, T.; Lee, S.H.; Kim, K.W.; Lim, H.K.; Choi, I.J.; et al. Exfoliation and Characterization of V_2Se_9 Atomic Crystals. *Nanomaterials* **2018**, *8*, 737. [CrossRef] [PubMed]

8. Lim, S.K.; Kang, S.C.; Yoo, T.J.; Lee, S.K.; Hwang, H.J.; Lee, B.H. Operation Mechanism of a MoS_2/BP Heterojunction FET. *Nanomaterials* **2018**, *8*, 797. [CrossRef] [PubMed]

9. Park, W.; Pak, Y.; Jang, H.Y.; Nam, J.H.; Kim, T.H.; Oh, S.; Choi, S.M.; Kim, Y.; Cho, B. Improvement of the Bias Stress Stability in 2D MoS_2 and WS_2 Transistors with a TiO_2 Interfacial Layer. *Nanomaterials* **2019**, *9*, 1155. [CrossRef] [PubMed]

10. Lee, J.S.; Park, C.S.; Kim, T.Y.; Kim, Y.S.; Kim, E.K. Characteristics of P-type Conduction in P-doped MoS_2 by Phosphorous Pentoxide during Chemical Vapor Deposition. *Nanomaterials* **2019**, *9*, 1278. [CrossRef] [PubMed]

11. Park, W.; Jang, H.Y.; Nam, J.H.; Kwon, J.; Cho, B.; Kim, Y. Artificial 2D van der Waals Synapse Devices via Interfacial Engineering for Neuromorphic Systems. *Nanomaterials* **2020**, *10*, 88. [CrossRef] [PubMed]

12. Ko, C. Reconfigurable Local Photoluminescence of Atomically-thin Semiconductors Via Ferroelectric-assisted Effects. *Nanomaterials* **2019**, *9*, 1620. [CrossRef] [PubMed]

13. Liang, S.; Hasan, M.N.; Seo, J.H. Direct Observation of Raman Spectra in Black Phosphorus under Uniaxial Strain Conditions. *Nanomaterials* **2019**, *9*, 566. [CrossRef] [PubMed]

14. Zhang, X.; Zhang, R.; Zheng, X.; Zhang, Y.; Zhang, X.; Deng, C.; Qin, S.; Yang, H. Interlayer Difference of Bilayer-Stacked MoS_2 Structure: Probing by Photoluminescence and Raman Spectroscopy. *Nanomaterials* **2019**, *9*, 796. [CrossRef] [PubMed]

15. Zhang, X.; Wang, J.; Xu, H.; Tan, H.; Ye, X. Preparation and Tribological Properties of WS Hexagonal Nanoplates and Nanoflowers. *Nanomaterials* **2019**, *9*, 840. [CrossRef] [PubMed]

 *nanomaterials*

Article

Artificial 2D van der Waals Synapse Devices via Interfacial Engineering for Neuromorphic Systems

Woojin Park [1], Hye Yeon Jang [1], Jae Hyeon Nam [1], Jung-Dae Kwon [2], Byungjin Cho [1,*] and Yonghun Kim [2,*]

[1] Department of Advanced Material Engineering, Chungbuk National University, Chungdae-ro 1, Seowon-Gu, Cheongju, Chungbuk 28644, Korea; wjpark@chungbuk.ac.kr (W.P.); hyjang0581@gmail.com (H.Y.J.); jhnam0714@gmail.com (J.H.N.)

[2] Materials Center for Energy Convergence, Surface Technology Division, Korea Institute of Materials Science (KIMS), 797 Changwondaero, Sungsan-gu, Changwon, Gyeongnam 51508, Korea; jdkwon@kims.re.kr

* Correspondence: bjcho@chungbuk.ac.kr (B.C.); kyhun09@kims.re.kr (Y.K.); Tel.: +82-(0)43-261-2417 (B.C.); +82-(0)55-280-3281 (Y.K.)

Received: 22 November 2019; Accepted: 31 December 2019; Published: 2 January 2020

Abstract: Despite extensive investigations of a wide variety of artificial synapse devices aimed at realizing a neuromorphic hardware system, the identification of a physical parameter that modulates synaptic plasticity is still required. In this context, a novel two-dimensional architecture consisting of a $NbSe_2/WSe_2/Nb_2O_5$ heterostructure placed on an $SiO_2/p+$ Si substrate was designed to overcome the limitations of the conventional silicon-based complementary metal-oxide semiconductor technology. $NbSe_2$, WSe_2, and Nb_2O_5 were used as the metal electrode, active channel, and conductance-modulating layer, respectively. Interestingly, it was found that the post-synaptic current was successfully modulated by the thickness of the interlayer Nb_2O_5, with a thicker interlayer inducing a higher synapse spike current and a stronger interaction in the sequential pulse mode. Introduction of the Nb_2O_5 interlayer can facilitate the realization of reliable and controllable synaptic devices for brain-inspired integrated neuromorphic systems.

Keywords: 2D heterostructure; WSe_2; $NbSe_2$; Nb_2O_5 interlayer; synapse device; neuromorphic system

---

1. Introduction

Continuous downscaling has stimulated the development of semiconductor technology for the last several decades, offering advantages, such as lower power consumption, higher integration, faster circuit operation, and reduced device cost per function. However, the side effects from continuous downscaling, to a size of less than 10 nm, limit the further development of the silicon semiconductor technology. This has motivated the exploration of novel computation systems beyond the conventional Von Neumann architecture that can overcome the downscaling limitations. Recently, due to the increasing need to implement sophisticated information processing system mimicking the human brain, the neuromorphic computing system has attracted a great deal of attention [1–5]. For the integrated neuromorphic systems, it is important to realize operations of complex and diverse functions implemented using a parallel architecture consisting of ~10^{11} neurons and ~10^{15} synapses. Additionally, the unit event should be simultaneously conducted using an extremely small amount of energy [6].

The artificial synapse device is considered to be an essential fundamental element for the emulation of biological neural networks [7]. The mechanism of operation for transmitting a spike input stimulus through the synapse can strengthen or weaken the synaptic weight, which is known as synaptic plasticity [8]. The synapse provides the functions of information processing and storage based on the spiking neural network. For this system, conventional solid-state electronics technology has been adopted for emulating the biological synapse function, in order to demonstrate a neuromorphic computing system [9]. In previous studies, conventional silicon-based complementary metal-oxide semiconductor (CMOS) technology was employed for demonstrating solid-state synapse devices, and a network consisting of 256 million configurable synapses and 1 million programmable spiking neurons was demonstrated [10]. The use of the 28-nm fully depleted silicon-on-insulator CMOS technology for 64k-synapse and 256-neuron architecture was also reported [11]. However, these CMOS-based devices are still unsuitable for realizing an artificial intelligence chip, because they cannot meet the requirements of higher integration density and lower power consumption. Si CMOS-based synapse device is based on the operation of complex logic circuits. This means that its power dissipation is essentially higher than that of other types, which is not satisfactory for emulating the biological synapse with an ultralow femtojoule energy consumption.

To eliminate the bottlenecks hindering the further development of neuromorphic computing systems, three-terminal artificial synaptic transistors, based on novel semiconductors have been studied to demonstrate synaptic functions. For instance, diverse semiconducting materials including carbon nanotubes, [12] nickelate, [13], and indium gallium zinc oxide (IGZO) [14,15] have been selected for the realization of synapse platforms. Meanwhile, two-dimensional (2D) transition dichalcogenides (TMDCs) are an intriguing nanomaterial layer for key elements of synaptic transistors due to their advantages of excellent intrinsic scalability, transparency, chemical robustness, and low power consumption [16–19]. In fact, several research groups have demonstrated the corresponding synaptic devices [20,21]. Meanwhile, a variety of oxide layers have been used as the conductance-tuning layers for synapse device applications. For example, phase change memory emulating synaptic behavior was demonstrated using a thin HfO_2 interface layer [22]. Additionally, Deswal et al. reported an NbO_x-based memristor, showing a gradual and continuous conductance change that is a prerequisite of a biological synapse device [23]. Nevertheless, it is still unclear what physical parameters can be used to precisely manipulate the synaptic functions. Thus, the use of a 2D heterostructure, combined with insulating oxide, can be an alternative approach for the development of energy-efficient artificial synapse devices.

In this work, we designed a vertically-stacked 2D metallic electrode $NbSe_2$/semiconductor WSe_2/interlayer Nb_2O_5 heterostructure placed on an Si/SiO_2 substrate with the back-gate configuration. Here, WSe_2 and Nb_2O_5 served as the active channel, and the conductance-tuning layer, respectively. Additionally, the $NbSe_2$ electrode can provide excellent transistor switching characteristics due to a sharp 2D interface and the absence of the metal-induced gap states [24,25]. The post-synaptic current behavior can be modulated precisely by adjusting the thickness of the Nb_2O_5 layer, with a thicker Nb_2O_5 interlayer providing higher synapse spike current and strong interaction in paired pulse facilitation testing modes. The charge trapping/detrapping mechanism at the Nb_2O_5 defect states based on an energy band model was proposed. The novel 2D architecture will pave the way toward extreme integration for the development of the massively parallel neuromorphic circuitry system.

2. Materials and Methods

2.1. CVD Synthesis of WSe$_2$ and NbSe$_2$

A selenium (Se)-based semiconducting channel based on WSe_2 and a metallic electrode based on $NbSe_2$ were synthesized using a simple two-step process. First, WO_3 and Nb_2O_5 thin films were individually deposited on an SiO_2/Si wafer. The thicknesses of the WO_3 and Nb_2O_5 thin films were ~3, and ~5 nm, respectively. This pre-deposited oxide layer on the wafer was directly loaded into the center of thermal furnace and vacuumed with a rotary pump system. Then, the thermal furnace was heated to the desired temperature (~1000 °C) under the flow of 5% hydrogen-balanced Ar gas (Ar/H_2), while a selenium powder source was sublimated by heating to 500 °C. After a 1-h selenization process, the furnace was naturally cooled down to room temperature.

2.2. Fabrication of 3-Terminal Synapse Device

A heavily doped p-type Si substrate with SiO_2 was cleaned by sonication in acetone, methanol, and iso-propyl alcohol (IPA) solution. To precisely tune the synaptic weight corresponding to the drain current, the charge trapping layer of the Nb_2O_5 thin film was deposited with different thicknesses using thermal evaporation. The thickness of Nb_2O_5 varied from 2.6 to 3.9 nm, as validated by the cross-sectional transmission electron microscopy (TEM) analysis. Then, the synthesized WSe_2 semiconducting channel was transferred onto an SiO_2/Si wafer using a poly(methyl methacrylate)-assisted transfer method and patterned using conventional photolithography. Finally, the $NbSe_2$ metallic electrode was transferred for the formation of the $NbSe_2/WSe_2$ van der Waals heterojunction, in order to minimize the contact resistance [24,25].

2.3. Electrical Characterization

Basic electrical characterizations were carried out using a Keithley 2636B source meter (Keithley Instruments, Solon, OH, USA). The amplitude of the applied synaptic pulse, used to generate an excitatory post-synaptic current (EPSC), was 20 V and its duration was varied from 2 to 10 s.

3. Results and Discussion

Figure 1a shows a schematic of a biological neural network consisting of synapses and neurons. The most important trait of brain-inspired devices is their capability for efficient data processing using an extremely small amount of power in the networks with an astronomical number of synapses and neurons. The parallel network means that processing and storage of information occur simultaneously and do not follow the von Neumann computing paradigm. Therefore, a high device integration density and low energy consumption are crucial for a neuromorphic system. The operation of transmitting a spike input stimulus is illustrated in Figure 1b. The interaction of the pre- and post-synaptic activities affects the long-lasting connection strength, and long-lasting plasticity is considered to be the key mechanism of basic neuromorphic computation. Figure 1c shows the back-gate configuration of the WSe_2 synapse transistor. The heavily-doped Si layer was used as the back-gate and $NbSe_2$ was used as the source/drain. The Nb_2O_5 interfacial layer allows the fine-tuning of the conductance of the WSe_2 transistor.

Figure 2a shows a schematic of the electrical measurements of the synapse device in the back-gate pulse system. Figure 2b shows the obtained cross-sectional high-resolution transmission electron microscopy images and the results of the energy-dispersive X-ray spectroscopy (EDS) analysis, thereby, clearly demonstrating the distinct film layers and sharp junction interfaces. The different stacking structures of WSe_2-$NbSe_2$, 2.6 nm Nb_2O_5-WSe_2-$NbSe_2$, and 3.9 nm Nb_2O_5-WSe_2-$NbSe_2$ were clearly observed and compared. The boundaries of each layer appeared to be atomically sharp and smooth without a significant interfacial gap. Five layers of $NbSe_2$ and three layers of WSe_2 were consistently observed for all of the samples, and the additional interfacial Nb_2O_5 layer was also clearly observed. The distributions of the W, Se, Nb, and O elements were obtained from the EDS elemental mapping images. The left panel of Figure 2b shows the WSe_2-$NbSe_2$ stack architecture without the Nb_2O_5 deposition. Since, both the $NbSe_2$ and Nb_2O_5 films contain Nb atoms, the two separate Nb layers were observed only in the samples with the Nb_2O_5 interfacial layer, verifying the existence of Nb_2O_5. The middle panel of Figure 2b shows the results for the sample with a 2.6 nm Nb_2O_5 layer. The right panel of Figure 2b shows the sample with a 3.6 nm Nb_2O_5 layer. Figure 2c shows that the Raman spectra obtained for the as-synthesized 2D films support the presence of 2D materials, such as WSe_2 and $NbSe_2$, demonstrating the successful synthesis of the 2D nanomaterials via the chemical vapor deposition (CVD) technique. The Raman spectra of WSe_2 and $NbSe_2$ clearly display the in-plane vibrational modes of W-Se and Nb-Se (E^1_{2g}: 250.3 and 243.2 cm^{-1}) and the out-of-plane vibrational modes that arise from the motion the Se atoms (A_{1g}: 258.5 and 230.6 cm^{-1} for WSe_2 and $NbSe_2$).

Furthermore, two distinct Raman peaks of WSe$_2$ and NbSe$_2$ with stacked device structure were also observed even after transfer process in Figure S1 in the Supplementary information.

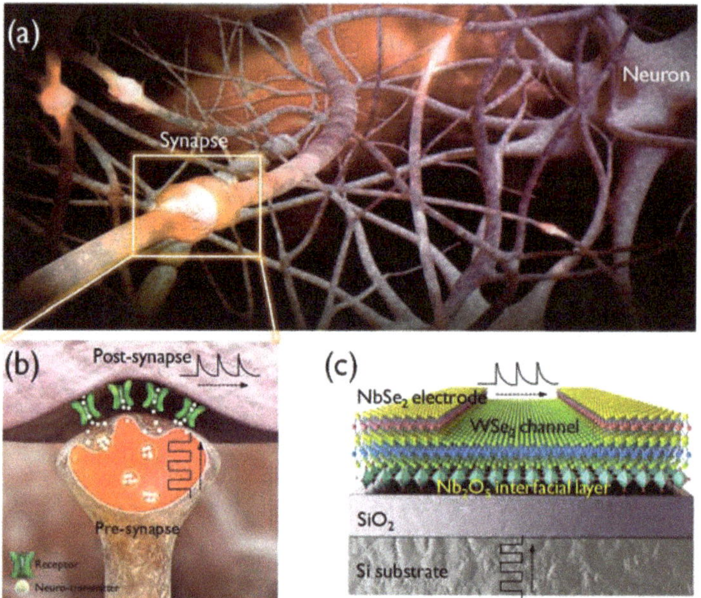

Figure 1. (**a**) Biological neural network consisting of synapses and neurons. (**b**) Operational mechanism of the transmission of an input stimulus from pre-synapse to post-synapse. (**c**) Artificial synapse transistor comprised by vertically stacked NbSe$_2$/WSe$_2$/Nb$_2$O$_5$/SiO$_2$/p+ Si, mimicking the function of bio synapse.

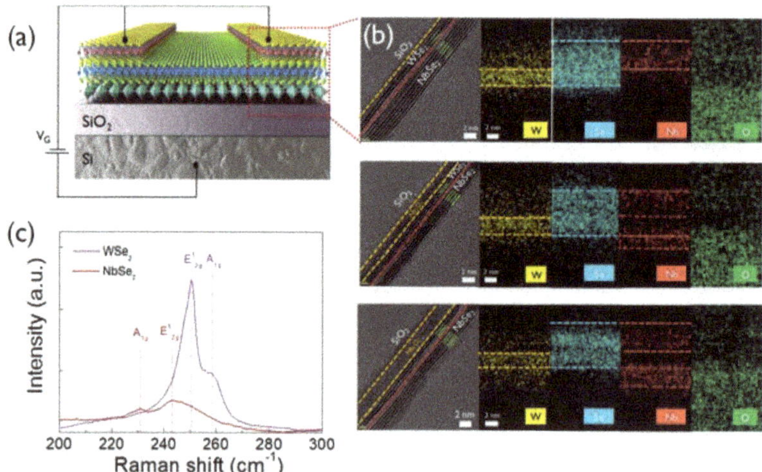

Figure 2. (**a**) Configuration scheme for the electrical measurements of the synapse transistor device. (**b**) Cross-sectional high-resolution transmission electron microscopy and energy-dispersive X-ray spectroscopy (EDS) elemental mapping images recorded from WSe$_2$-NbSe$_2$, 2.6 nm Nb$_2$O$_5$-WSe$_2$-NbSe$_2$ and, 3.9 nm Nb$_2$O$_5$-WSe$_2$-NbSe$_2$ (**c**) Raman spectra for WSe$_2$, and NbSe$_2$ that serve as the active channel and metallic electrode, respectively.

To compare the transfer characteristics of the WSe$_2$-NbSe$_2$ van der Waals hetero-junction devices with different Nb$_2$O$_5$ thickness, DC-mode-based double sweep measurements were performed, as shown in Figure 3a. The double sweep curves of the 2D heterojunction devices were obtained under varying values of able V$_{BG}$ in the range from 10 to −20 V at a fixed drain voltage of −5 V. The WSe$_2$-based transistor showed typical p-type unipolar behavior, with a counterclockwise hysteresis loop, that may be ascribed to the confinement of the hole charges in the trap states induced by the Nb$_2$O$_5$ interlayer [26]. Additionally, the repeatability test of DC transfer double sweep curves, with different Nb$_2$O$_5$ thicknesses, were also shown in Figure S2 in Supplementary information. We also investigated the statistical distribution of the hysteresis window voltages, in order to validate the reliability of the data corresponding to the hysteresis behavior (Figure 3b). The average values of the hysteresis voltage for each device were measured to be ~5, 7, and 11 V, respectively. The value of the error bar was almost same for all of the devices. Thus, it is clear that a thicker Nb$_2$O$_5$ interlayer gives rise to a larger hysteresis window. The dependence of DC sweep speed on transfer curves was also depicted in Figure S3 in Supplementary information.

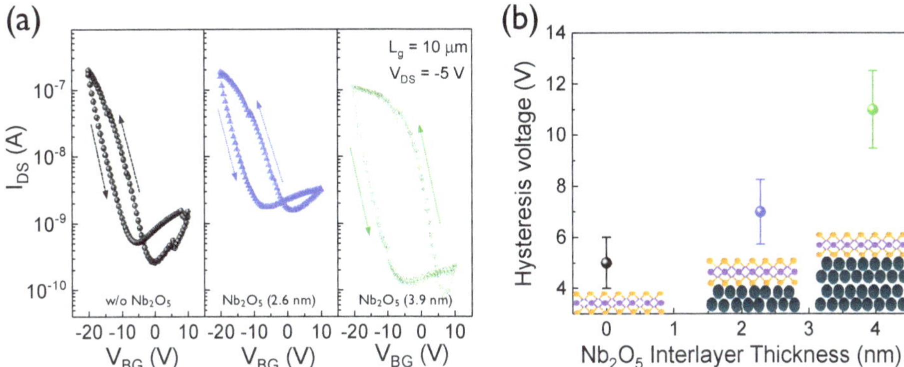

Figure 3. (a) Hysteresis behaviors of the two-dimensional (2D) WSe$_2$-NbSe$_2$ hetero-structure transistor devices with different Nb$_2$O$_5$ interlayer thickness. (b) Hysteresis window voltage as a function of the Nb$_2$O$_5$ interlayer thickness for the 2D WSe$_2$-NbSe$_2$ devices.

To elucidate the origin of the hysteresis of the 2D heterostructure transistors, the corresponding energy band model was proposed (Figure 4). We previously reported the positive effect of the combination of WSe$_2$-NbSe$_2$ with reduced contact barrier [24,25]. The conventional Richardson-Schottky equation was employed to calculate Schottky barrier,

$$I_{DS} = AA^*T^2 \exp\left[-\frac{\left(\Phi_B - \sqrt{q^3 V/4\pi\varepsilon_0\varepsilon_r d}\right)}{k_b T}\right] \tag{1}$$

where A is the contact area, A^* is the effective Richardson constant, T is the temperature, Φ_B is the Schottky barrier height, q is the electron charge, V is the applied forward bias, ε_0 and ε_r are the permittivity of the vacuum and the oxide layer, respectively, d is the width of the interface barrier, and K_b is the Boltzmann constant. It was mentioned in the references that Schottky barrier at WSe$_2$-NbSe$_2$ contact is significantly lower than that at WSe$_2$-metal(Pd) contact due to Fermi-level de-pinning. Therefore, the 2D WSe$_2$-NbSe$_2$ combination can be an excellent candidate for the fabrication of an energy-efficient low-power synaptic transistor, due to its low contact resistance. Recently, the new methodology for universal 2D material was reported to obtain Schottky barrier, suggesting more accurate calculation [27]. Holes are known to be the major carriers in both the semiconductor channel WSe$_2$ and the metallic source/drain electrode NbSe$_2$. Thus, only the hole charge transport was considered in our proposed switching model. As shown in Figure 4a, the negative voltage applied to the back gate electrode (p +Si) shifts the corresponding Fermi level upward, accumulating hole charge near the Nb_2O_5-corresponding defect states. Under a negative gate bias, holes can be easily trapped in the defect states within the Nb_2O_5 interlayer, depleting the carriers in the WSe_2 and leading to a decrease in the drain current. Meanwhile, when a positive voltage is applied to the gate, the Fermi level shifts downward, depleting the trapped holes in the Nb_2O_5 defects (Figure 4b). Simply put, the trapped holes will be released across the Nb_2O_5-WSe_2 interface, leading to an increase in the drain current. Indeed, we experimentally proved that the amount of the trapped hole carriers is controlled by the Nb_2O_5 thickness.

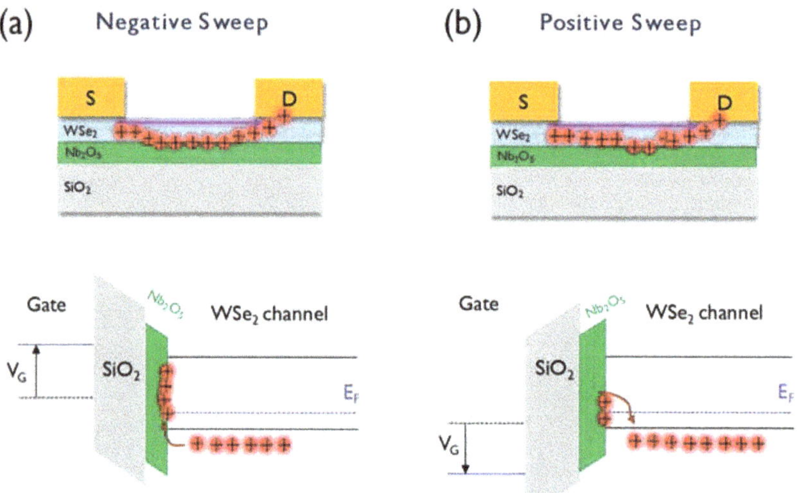

Figure 4. (a) Device operation scheme and energy band model of the 2D heterostructure transistor device for describing the trapping behavior of the hole carriers at the negative gate bias condition; (**b**) device operation scheme and energy band model of the 2D heterostructure transistor device, corresponding to the process of the release of the trapped hole carriers at a positive gate bias condition.

To characterize the pulse response of the 2D heterostructure devices, we monitored the spike current response to the gate voltage pulses with the amplitude and duration time of 20 V, and 2 s, respectively (Figure 5a). In neuroscience, it is important to transfer electrical or chemical signal from pre-synapse to post-synapse. This is usually caused by the flow of positively charged ions. EPSC can be generated by the action of ions or electron flow in the neuromorphic system. The gate voltage for the EPSC was fixed at −20 V to give a fair comparison for each case. EPSC reaches the maximum value and then decays back to the initial current state. Interestingly, the spike was generated, even in the reference device without Nb_2O_5 layer. This might be because of the unintentional charge trap sites, which exist at the diverse interfaces (WSe_2-$NbSe_2$ and SiO_2-WSe_2). Our result showed that the peak values increase with the increasing thickness of the inserted Nb_2O_5 interlayer. Higher voltage pulses required long decay time to restore the synapse device to the initial current state, leading to stronger nonvolatile properties. The duration time of the pulse voltage, that is applied to the devices also affected the peak EPSC (Figure 5b). A longer pulse duration resulted in a higher peak EPSC. In a biological neural network, paired pulse facilitation is an important synapse parameter for determining synaptic plasticity, that is responsible for learning and memory processes [28]. As shown in Figure 5c, paired pulse facilitation is the phenomenon where the EPSC stimulated by the second spike is enhanced when the first spike is closely followed by the second spike [29,30]. Such essential synapse behavior can be emulated using our 2D heterostructure transistor. Figure 5d shows the interaction of two sequential spikes for all of the devices. The interval time between the applied pulses was 2 s. The interaction between the output spike current of the control device without Nb_2O_5 was not observed, indicating the negligible synaptic weight modulation property. Meanwhile, the introduction of the Nb_2O_5 layer strengthened the interaction of the two spikes; more specifically, a thicker Nb_2O_5 interlayer induced a much greater current change at the second pulse mode. Such a strong tuning ability of the synaptic weight enhances the electrical plasticity of the artificial synapse device, and may improve the intelligence of the integrated neuromorphic system [12].

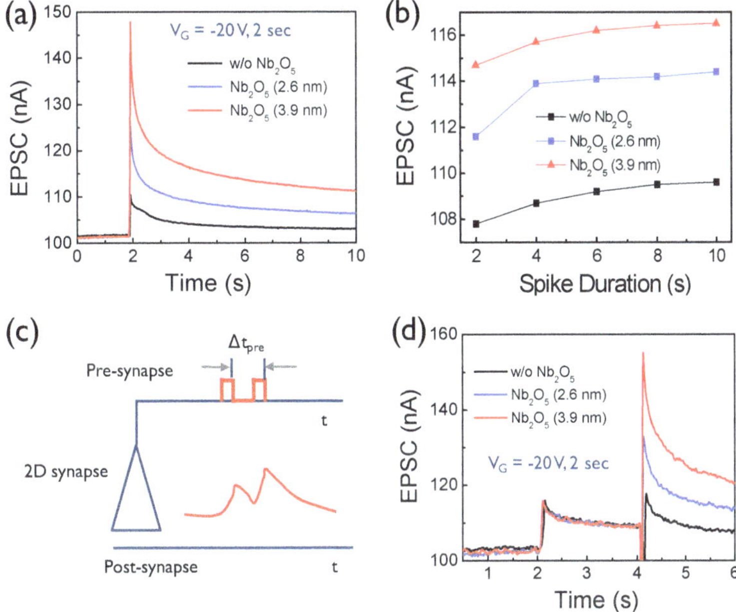

Figure 5. (**a**) Comparison of the excitatory post-synaptic current (EPSC) behavior of the 2D heterostructure devices with different Nb_2O_5 interlayer thicknesses. (**b**) Comparison of the EPSC data as a function of the spike duration time for the different 2D heterostructure devices. (**c**) Operation scheme of the synapse circuit for describing paired pulse facilitation that is stimulated by the application of two sequential pulses. (**d**) Comparison of paired pulse facilitation behavior for the different 2D heterostructure devices.

4. Conclusions

We demonstrated controllable synaptic plasticity with the WSe_2/Nb_2O_5 heterostructure in the WSe_2 back-gate device. The Nb_2O_5 layer served as the conductance-modifying layer and enabled precise modulation of the conductive states and their dynamic change. Essential synaptic functions (EPSC and paired pulse facilitation) were investigated in the WSe_2/Nb_2O_5 heterostructure devices. In particular, the optimized thickness of the Nb_2O_5 layer strengthened the interaction in the synaptic weight, showing the largest post-synapse current. Thus, the facile one-step Nb_2O_5 layer deposition process, demonstrated in this work, is an effective approach for the realization of controllable synaptic devices.

Supplementary Materials: The following are available online at http://www.mdpi.com/2079-4991/10/1/88/s1, Figure S1: Raman spectrum of WSe_2 and $NbSe_2$ after transfer process, Figure S2: Repeatability test of DC transfer double sweep curves with different Nb_2O_5 thickness, Figure S3: The dependence of DC sweep speed on transfer curves.

Author Contributions: B.C. and Y.K. designed and conducted the experiments. H.Y.J. and J.H.N. and J.-D.K. supported the execution of the experiments and the data analysis. W.P., B.C., and Y.K. wrote the paper. B.C., and Y.K. supported and guided the experiments and analysis of results. Y.K. conceived the project and supervised the writing of the paper. All authors have read and agreed to the published version of the manuscript.

Funding: This work was supported by the National Research Foundation of Korea (NRF) grant funded by the Korean government (MSIT; Ministry of Science and ICT) (No. 2017R1C1B1005076) and Fundamental Research Program (No. PNK6990) of the Korea Institute of Materials Science (KIMS). This research was also financially supported by the Ministry of Trade, Industry and Energy (MOTIE) and Korea Institute for Advancement of Technology (KIAT) through the National Innovation Cluster R & D program (P0006704_Development of energy saving advanced parts).

Conflicts of Interest: The authors declare no conflict of interest.

References

1. Kim, S.; Yoon, J.; Kim, H.D.; Choi, S.J. Carbon Nanotube Synaptic Transistor Network for Pattern Recognition. *ACS Appl. Mater. Interfaces* **2015**, *7*, 25479–25486. [CrossRef]
2. Fuller, E.J.; Gabaly, F.E.; Léonard, F.; Agarwal, S.; Plimpton, S.J.; Jacobs-Gedrim, R.B.; James, C.D.; Marinella, M.J.; Talin, A.A. Li-Ion Synaptic Transistor for Low Power Analog Computing. *Adv. Mater.* **2017**, *29*, 1604310. [CrossRef] [PubMed]
3. Van De Burgt, Y.; Lubberman, E.; Fuller, E.J.; Keene, S.T.; Faria, G.C.; Agarwal, S.; Marinella, M.J.; Alec Talin, A.; Salleo, A. A non-volatile organic electrochemical device as a low-voltage artificial synapse for neuromorphic computing. *Nat. Mater.* **2017**, *16*, 414–418. [CrossRef] [PubMed]
4. Choi, S.; Tan, S.H.; Li, Z.; Kim, Y.; Choi, C.; Chen, P.Y.; Yeon, H.; Yu, S.; Kim, J. SiGe epitaxial memory for neuromorphic computing with reproducible high performance based on engineered dislocations. *Nat. Mater.* **2018**, *17*, 335–340. [CrossRef] [PubMed]
5. Kim, H.; Hwang, S.; Park, J.; Park, B.G. Silicon synaptic transistor for hardware-based spiking neural network and neuromorphic system. *Nanotechnology* **2017**, *28*, 405202. [CrossRef] [PubMed]
6. Cao, Q.L.; Yan, X.X.; Luo, X.G.; Garey, L.J. Prenatal development of parvalbumin immunoreactivity in the human striate cortex. *Cereb. Cortex* **1996**, *6*, 620–630. [CrossRef] [PubMed]
7. Li, G.L.; Keen, E.; Andor-Ardó, D.; Hudspeth, A.J.; Von Gersdorff, H. The unitary event underlying multiquantal EPSCs at a hair cell's ribbon synapse. *J. Neurosci.* **2009**, *29*, 7558–7568. [CrossRef]
8. Kauer, J.A.; Malenka, R.C. Synaptic plasticity and addiction. *Nat. Rev. Neurosci.* **2007**, *8*, 844–858. [CrossRef]
9. Shi, R.Z.; Horiuchi, T. A summating, exponentially-decaying CMOS synapse for spiking neural systems. In Proceedings of the Advances in Neural Information Processing Systems. *Proc. Adv. Neural Inf. Process. Syst.* **2004**, *16*, 1003–1010.
10. Merolla, P.A.; Arthur, J.V.; Alvarez-icaza, R.; Cassidy, A.S.; Sawada, J.; Akopyan, F.; Jackson, B.L.; Imam, N.; Guo, C.; Nakamura, Y.; et al. A million spiking-neuron integrated circuit with a scalable communication network and interface. *Science* **2014**, *345*, 668–673. [CrossRef]
11. Frenkel, C.; Lefebvre, M.; Legat, J.D.; Bol, D. A 0.086-mm^2 12.7-pJ/SOP 64k-Synapse 256-Neuron Online-Learning Digital Spiking Neuromorphic Processor in 28-nm CMOS. *IEEE Trans. Biomed. Circuits Syst.* **2019**, *13*, 145–158. [PubMed]
12. Kim, S.; Choi, B.; Lim, M.; Yoon, J.; Lee, J.; Kim, H.D.; Choi, S.J. Pattern Recognition Using Carbon Nanotube Synaptic Transistors with an Adjustable Weight Update Protocol. *ACS Nano* **2017**, *11*, 2814–2822. [CrossRef] [PubMed]
13. Shi, J.; Ha, S.D.; Zhou, Y.; Schoofs, F.; Ramanathan, S. A correlated nickelate synaptic transistor. *Nat. Commun.* **2013**, *4*, 2676. [CrossRef] [PubMed]
14. Pillai, P.B.; De Souza, M.M. Nanoionics-based three-terminal synaptic device using zinc oxide. *ACS Appl. Mater. Interfaces* **2017**, *9*, 1609–1618. [CrossRef]
15. Shao, F.; Yang, Y.; Zhu, L.Q.; Feng, P.; Wan, Q. Oxide-based Synaptic Transistors Gated by Sol-Gel Silica Electrolytes. *ACS Appl. Mater. Interfaces* **2016**, *8*, 3050–3055. [CrossRef]
16. Radisavljevic, B.; Radenovic, A.; Brivio, J.; Giacometti, V.; Kis, A. Single-layer MoS$_2$ transistors. *Nat. Nanotechnol.* **2011**, *6*, 147–150. [CrossRef]
17. Yin, Z.; Li, H.; Li, H.; Jiang, L.; Shi, Y.; Sun, Y.; Lu, G.; Zhang, Q.; Chen, X.; Zhang, H. Single-Layer MoS$_2$ Phototransistors. *ACS Nano* **2012**, *6*, 74–80. [CrossRef]

18. Bertolazzi, S.; Brivio, J.; Kis, A. Stretching and breaking of ultrathin MoS$_2$. *ACS Nano* **2011**, *5*, 9703–9709. [CrossRef]

19. Salvatore, G.A.; Münzenrieder, N.; Barraud, C.; Petti, L.; Zysset, C.; Büthe, L.; Ensslin, K.; Tröster, G. Fabrication and transfer of flexible few-layers MoS$_2$ thin film transistors to any arbitrary substrate. *ACS Nano* **2013**, *7*, 8809–8815. [CrossRef]

20. Zhu, J.; Yang, Y.; Jia, R.; Liang, Z.; Zhu, W.; Rehman, Z.U.; Bao, L.; Zhang, X.; Cai, Y.; Song, L.; et al. Ion Gated Synaptic Transistors Based on 2D van der Waals Crystals with Tunable Diffusive Dynamics. *Adv. Mater.* **2018**, *30*, 1800195. [CrossRef]

21. Sangwan, V.K.; Lee, H.S.; Bergeron, H.; Balla, I.; Beck, M.E.; Chen, K.S.; Hersam, M.C. Multi-terminal memtransistors from polycrystalline monolayer molybdenum disulfide. *Nature* **2018**, *554*, 500–504. [CrossRef] [PubMed]

22. Suri, M.; Bichler, O.; Hubert, Q.; Perniola, L.; Sousa, V.; Jahan, C.; Vuillaume, D.; Gamrat, C.; Desalvo, B. Addition of HfO$_2$ interface layer for improved synaptic performance of phase change memory (PCM) devices. *Solid-State Electron.* **2013**, *79*, 227–232. [CrossRef]

23. Deswal, S.; Kumar, A.; Kumar, A. NbOx based memristor as artificial synapse emulating short term plasticity. *AIP Adv.* **2019**, *9*, 095022. [CrossRef]

24. Kim, A.R.; Kim, Y.; Nam, J.; Chung, H.S.; Kim, D.J.; Kwon, J.D.; Park, S.W.; Park, J.; Choi, S.Y.; Lee, B.H.; et al. Alloyed 2D Metal-Semiconductor Atomic Layer Junctions. *Nano Lett.* **2016**, *16*, 1890–1895. [CrossRef]

25. Kim, Y.; Kim, A.R.; Yang, J.H.; Chang, K.E.; Kwon, J.D.; Choi, S.Y.; Park, J.; Lee, K.E.; Kim, D.H.; Choi, S.M.; et al. Alloyed 2D Metal-Semiconductor Heterojunctions: Origin of Interface States Reduction and Schottky Barrier Lowering. *Nano Lett.* **2016**, *16*, 5928–5933. [CrossRef]

26. Wang, J.C.; Kao, C.H.; Wu, C.H.; Lin, C.F.; Lin, C.J. Nb$_2$O$_5$ and Ti-doped Nb$_2$O$_5$ charge trapping nano-layers applied in flash memory. *Nanomaterials* **2018**, *8*, 799. [CrossRef]

27. Ang, Y.S.; Yang, H.Y.; Ang, L.K. Universal Scaling Laws in Schottky Heterostructures Based on Two-Dimensional Materials. *Phys. Rev. Lett.* **2018**, *121*, 56802. [CrossRef]

28. Zhou, J.; Wan, C.; Zhu, L.; Shi, Y.; Wan, Q. Synaptic behaviors mimicked in flexible oxide-based transistors on plastic substrates. *IEEE Electron Device Lett.* **2013**, *34*, 1433–1435. [CrossRef]

29. John, R.A.; Ko, J.; Kulkarni, M.R.; Tiwari, N.; Chien, N.A.; Ing, N.G.; Leong, W.L.; Mathews, N. Flexible Ionic-Electronic Hybrid Oxide Synaptic TFTs with Programmable Dynamic Plasticity for Brain-Inspired Neuromorphic Computing. *Small* **2017**, *13*, 15–23. [CrossRef]

30. Sun, L.; Zhang, Y.; Hwang, G.; Jiang, J.; Kim, D.; Eshete, Y.A.; Zhao, R.; Yang, H. Synaptic Computation Enabled by Joule Heating of Single-Layered Semiconductors for Sound Localization. *Nano Lett.* **2018**, *18*, 3229–3234. [CrossRef]

 *nanomaterials*

Article

Reconfigurable Local Photoluminescence of Atomically-Thin Semiconductors via Ferroelectric-Assisted Effects

Changhyun Ko [1,2]

[1] Department of Applied Physics, College of Engineering, Sookmyung Women's University, Seoul 04310, Korea; cko@sookmyung.ac.kr; Tel.: +82-2-6325-3184
[2] Institute of Advanced Materials and Systems, Sookmyung Women's University, Seoul 04310, Korea

Received: 4 October 2019; Accepted: 11 November 2019; Published: 15 November 2019

Abstract: Combining a pair of materials of different structural dimensions and functional properties into a hybrid material system may realize unprecedented multi-functional device applications. Especially, two-dimensional (2D) materials are suitable for being incorporated into the heterostructures due to their colossal area-to-volume ratio, excellent flexibility, and high sensitivity to interfacial and surface interactions. Semiconducting molybdenum disulfide (MoS_2), one of the well-studied layered materials, has a direct band gap as one molecular layer and hence, is expected to be one of the promising key materials for next-generation optoelectronics. Here, using lateral 2D/3D heterostructures composed of MoS_2 monolayers and nanoscale inorganic ferroelectric thin films, reversibly tunable photoluminescence has been demonstrated at the microscale to be over 200% upon ferroelectric polarization reversal by using nanoscale conductive atomic force microscopy tips. Also, significant ferroelectric-assisted modulation in electrical properties has been achieved from field-effect transistor devices based on the 2D/3D heterostructrues. Moreover, it was also shown that the MoS_2 monolayer can be an effective electric field barrier in spite of its sub-nanometer thickness. These results would be of close relevance to exploring novel applications in the fields of optoelectronics and sensor technology.

Keywords: transition metal dichalcogenides; molybdenum disulfide; two-dimensional materials; ferroelectrics; photoluminescence

---

1. Introduction

Semiconducting layered transition metal dichalcogenides (TMDs) have been studied widely due to their strikingly interesting electronic and optoelectronic aspects unveiled in the two-dimensional (2D) limit including thickness-dependent bandgap, indirect-to-direct transitions of band structures, deeply-bound excitonic states, environment-sensitive characteristics, and so on [1–5]. Beyond fundamental interests, diverse device applications have also been realized from elementary field-effect transistors (FETs) and light-emitting diodes to complicated microprocessor structures as well as functional device components which enable memory, sensing, and resistive switching effects [6–10]. This ever-rising eagerness for the 2D TMD-based applications stems from a strong demand to replace Si-based components exclusively employed in the modern nano-electronics with the class of 2D semiconductors to overcome the fundamental scaling limitations, boost up integration density, and more innovatively, invent novel device platforms such as ultrathin flexible electronics [11].

However, despite the continual progress in this field, it is still challenging to incorporate 2D semiconductors into commercial products since material synthesis and device fabrication processes with 2D materials are currently not very cost-efficient and limited in scalability. Also, their operational device parameters such as carrier mobility and power consumption are expected to be inferior, out of laboratory, to those of contemporary devices which have been optimized for several decades [12]. Therefore, as promising alternative routes, diverse approaches have been made to assemble 2D semiconductors and a variety of 3D thin films structures which are very suitable for contemporary device fabrication infrastructure including classical dielectrics, semiconducting layered crystals, strongly correlated oxides, ferromagnetic materials, and ferroelectric (FE) thin epitaxies [8,12–18]. In these 2D/3D assemblies, in addition to size scaling benefits naturally given by introducing 2D materials, interfacial interference effects or thin films' functional attributes or both can be instilled into the 2D counterpart leading to achieving unique synergetic functionalities [12].

FE material systems where the electric polarization is built up spontaneously and can be flipped to the opposite direction abruptly by electrical stimulus have been considered intensively for various applications such as nonvolatile memory devices, photodetectors, water splitting, and photocatalysts [19–23]. Most of all, the ferroelectric FET (FeFET) where a dielectric layer in the conventional FET is replaced by a FE thin film has been considered as a promising candidate of next-generation memory with an advantage of nondestructive readout operation [19]. In the FeFET, the channel current can be modulated by ferroelectric gating. Due to the spontaneous polarization created in the FE thin film, the current can be maintained even after the removal of the gate voltage. More importantly, the FeFET structure can be realized in 2D/3D heterostructures simply by positioning 2D semiconductors as current channels on FE thin films [8,19]. Previously, the author of this paper and colleagues fabricated high-performance 2D/3D FeFET memory devices based on 2D MoS_2 and WSe_2 layers prepared by mechanical exfoliation from corresponding single crystals and lead zirconate titanate (PZT) epitaxial FE thin films [8]. More interestingly, the reversible nonvolatile photoluminescence (PL) modulation was also observed from the monolayer MoS_2 (ML-MoS_2) on the PZT thin film [8]. More recently, the ferroelectric control of PL was also demonstrated on other types of 2D TMDs, mechanically-exfoliated ML-$MoSe_2$ and ML-WSe_2 on the domain-engineered lithium niobate surface by B. Wen. et al. [24]. Also, M. Si. et al. reported fully-layered FeFET structures where 2D MoS_2 layers are interfaced with ferrielectric $CuInP_2S_6$ layered crystals [25].

As another type of FE-based device, the ferroelectric tunnel junction (FTJ) where a FE thin film is typically sandwiched by two metallic electrodes shows out-of-plane current ON/OFF switching via the change in band structure upon polarization reversal [26,27]. T. Li et al. demonstrated FTJs in the use of conductive atomic force microscopy (CAFM) technique on the 2D MoS_2/thin film $BaTiO_3$ heterostructures in which the top electrodes are few-layer MoS_2 layers grown by chemical vapor deposition (CVD) [26]. More recently, A. Lipatov et al., by local access to nanoscale FE domains via CAFM, realized programmable 1D current paths on CVD-grown ML-MoS_2 flakes using ferroelectric effects [27]. However, in this case, the in-plane current modulation shows the opposite trend to that observed in the typical FeFET with respect to the polarization direction. Although the authors argue that the in-plane current and out-of-plane current measured in the FTJ geometry on the identical devices should be correlated somehow, to elucidate mechanisms clearly, it would be necessary to investigate the FE effects into both device characteristics and optical properties [2,8,24,26,27]. Further, in the case of CVD-grown TMD MLs, inherent defective structures usually exist and hence, in comparison to the mechanically exfoliated MLs, the defect-sensitive characteristics should be carefully considered to fully understand the modulation of electrical and optical properties of TMD MLs driven by ferroelectric effects [28–31].

In this work, exploiting heterostructures of ML-MoS_2/FE thin epitaxy, the reversible modulation of nonvolatile PL, has been demonstrated at the microscale via ferroelectric-assisted effects. The electric field was applied directly through the heterostructures with both atomic formic microscopy (AFM) probe tips and electrical back-gated devices. Moreover, the polarization-induced manipulation in electrical characteristics was also realized on the FET devices based on the heterostructures along with simultaneous control of optical properties of the ML-MoS_2 channels. Lastly, in the same ML-MoS_2/FE geometry, I have also found that ML-MoS_2 sheets, of only sub-nanometer thickness, may play a role of an electric field barrier properly. The ML-MoS_2 layers were grown by the CVD method. Considering the recent progress in the CVD growth of 2D materials, this work would be relevant to the scalable and controllable fabrication of novel multi-dimensional devices [32]. As the FE components, two different inorganic FE materials were employed: PZT and $BiFeO_3$ (BFO) epitaxial thin films. The former has been studied widely for device applications due to their ultrafast dipole dynamics, good thermal and mechanical stability, and high dielectric breakdown limit [8,19,33]. Moreover, the latter produces the stronger polarization field which may lead to better device functionality [34,35].

2. Materials and Methods

2.1. MoS$_2$ Monolayer Growth

Triangular-shaped ML-MoS$_2$ flakes were grown on 100-nm-thick SiO$_2$/Si substrates via CVD after cleansing the substrates with Piranha solution and deionized (DI) water in a series. The ML-MoS$_2$ growth was carried out with the substrates face-down on an alumina crucible containing 3 mg of MoO$_3$ powder while the other crucible with S powder was located closer to the gas source than that with MoO$_3$ source. Initially, the furnace tube was purged by flowing ultrahigh purity N$_2$ gas at a flow rate of 500 SCCM (standard cubic centimeters per minute) for 10 min. Then, the heating process was performed in two steps: (1) The system was heated up to 300°C for ~10 min flowing N$_2$ at 100 SCCM; and (2) the temperature was increased up to 700 °C above the boiling temperature of S (~450 °C) within 15 min with N$_2$ gas flow of 5 SCCM and these conditions were sustained for 3 min. The furnace power was then shut down. Once the temperature reached 680 °C, the furnace was slightly opened and at 550 °C, the growth tube was detached entirely from the furnace to quench the samples. S vapor was supplied to the samples continuously even during the cooling step at a flow rate in the range of 2 to 5 SCCM to preserve the sample quality.

2.2. Ferroelectric Thin Film Synthesis and Heterostructure Fabrication

In this study, PZT (Pb[Zr$_{0.2}$Ti$_{0.8}$]O$_3$) and BFO (BiFeO$_3$) thin films were employed as the FE components of the ML-MoS$_2$/FE heterostructures. PZT (or BFO) thin films were grown epitaxially with a thickness of ~500 nm (or ~100 nm) on (001) STO (SrTiO$_3$) single-crystal substrates coated with SRO (SrRuO$_3$) layers by pulsed-laser deposition (PLD) utilizing a KrF laser (wavelength: ~248 nm) under a substrate temperature of 630 °C (or 690 °C), respectively. The buffer layer SRO was deposited at 740 °C also by PLD. During the deposition, a small amount of O$_2$ gas was supplied into the chamber maintaining the total pressure at ~100 mTorr. Subsequent to the deposition processes, the films were cooled down to room temperature at an O$_2$ pressure of 500 Torr with a cooling rate of 5 °C/min. The various thin film evaluations showed that the FE films used in this work have high quality elsewhere [8,29]. Then, the ML-MoS$_2$/FE heterostructures were constructed by transferring the as-grown MoS$_2$ flakes from SiO$_2$/Si substrates onto the FE thin films surface using polydimethylsiloxane (PDMS) films. The ML-MoS$_2$ flakes were detached from the SiO$_2$/Si substrates by wet etching with KOH solution. Also, before the transfer process, the surfaces of FE thin films were cleaned properly by gentle oxygen plasma etching.

2.3. Characterization of MoS$_2$ Monolayers and Heterostructures

Micro-PL and Raman experiments were conducted using objective lenses on a Renishaw micro-Raman/PL system (Gloucestershire, UK) operated with an excitation laser of ~488 nm wavelength in ambient conditions. The laser power was set properly in the range from 0.1 to 1 µW depending on the laser scan time to avoid any damage on the ML-MoS$_2$ sheets as well as FE thin films. The focused laser beam covers an area of ~6 µm^2. Out-of-plane piezoresponse force microscopy (PFM) was carried out on Veeco Multiprobe system (Plainview, NY, USA) to visualize the domain structure which is controllable by an electric field at the sub-microscale in air. FE domains were manipulated by applying a voltage in the range of −12 V to +12 V through conventional conductive AFM tips. Contact mode

AFM was also performed simultaneously to obtain surface topography images of ML-MoS$_2$ flakes on FE thin films.

2.4. Device Fabrication and Electrical Characterization

Conventional electron beam lithography was processed to lay out the electrode patterns. Ti/Au metal contacts with 10 nm/70 nm thicknesses were fabricated by e-beam evaporation and subsequent lift-off process. Back-gated FET measurements were conducted on a standard probe station equipped with two Keithley 617 programmable electrometers (Cleveland, OH, USA) which were designed to supply voltage to devices and to measure the source-drain and leakage currents separately in the ambient conditions.

3. Results and Discussion

3.1. 2D MoS$_2$/FE Heterostructure Preparation

Figure 1a shows a group of MoS$_2$ flakes grown on a SiO$_2$/Si substrate with a triangular shape, as typically observed on the family of CVD-grown TMD MLs. By the wet-transfer method, the MoS$_2$ sheets were micro-positioned on the clean surface area of ~500-nm-thick PZT thin films solidly, as shown in the optical microscopy image of Figure 1b. As summarized in Figure 1c,d, the optical characterization was conducted to determine the layer number of the CVD-grown MoS$_2$ flakes as well as to evaluate their quality. In Figure 1c, the PL spectrum of the MoS$_2$ flake clearly indicates the direct band-gap nature with the strong PL peak at ~1.84 eV verifying that the MoS$_2$ flakes used in this work are MLs [1,2,5,8]. Based on the Raman spectrum in Figure 1d, ML-MoS$_2$ sheets on PZT surface are not strained or damaged significantly through the wet-transfer process. Optical characterization was also performed on the ML-MoS$_2$/BFO heterostructures (see Supplementary Figure S1).

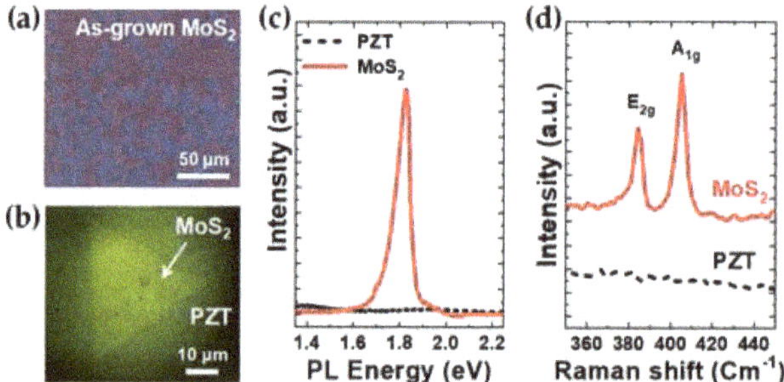

Figure 1. ML-MoS$_2$/PZT heterostructure preparation and optical characterization. Optical microscopy images of (**a**) as-grown ML-MoS$_2$ flakes on a SiO$_2$/Si substrate and (**b**) a representative ML-MoS$_2$ flake wet-transferred on a PZT thin film surface. (**c**) PL and (**d**) Raman spectra measured from the ML-MoS$_2$ flake displayed in (**a**) and the bare PZT surface as a reference.

3.2. Reversible Ferroelectric Modulation of Local Photoluminescence

Many semiconducting members of the TMD group, such as MoS$_2$, WS$_2$, WSe$_2$, MoSe$_2$, etc., show direct band gaps with a single molecular layer while indirect band gap structures are observed in their bulk counterparts [1,5,8,24]. Therefore, through the radiative exciton recombination, strong PL emission is observed in the direct-gap MLs. Further, the optoelectronic characteristics are strongly affected by the existence of quasiparticles including neutral excitons (X) of e-h pairs and charged excitons, also called trions, (X$^-$ or X$^+$) of e-e-h or e-h-h complexes, respectively, even at room temperature in contrast with conventional semiconductors. Therefore, the PL emissions of ML-MoS$_2$ can be modulated in terms of peak intensity and position, controlling the concentrations and types of the species. Up to now, numerous approaches have been demonstrated for altering the PL emissions:

electrostatic charging, chemical treatment, physisorption, point defect formation, substitutional doping, and so on [2,5,24,31,36,37]. In this work, using spontaneous polarization of FE materials, the densities of excitons and trions in ML-MoS$_2$ were controlled in a nonvolatile way and eventually, optoelectronic memory effects were realized in two-different ways: (1) local gating through conductive AFM tips, called the poling process, and (2) electrical back-gating with patterned metal electrodes.

Figure 2a,b shows schematically how the poling process is performed by applying poling voltage (V_P) using an AFM tip across a PZT thin film sandwiched by a ML-MoS$_2$ sheet and a metallic SRO layer which play roles as electrical pads on each side. When the V_P is large enough to induce polarization reversal above the coercive field of the PZT thin film, the up-polarized (P$_\uparrow$) and down-polarized (P$_\downarrow$) states are built up with the negative and positive V_P, respectively, and the polarization states are maintained even after the detachment of the AFM tip due to spontaneous polarization. Based on the schematics, it is expected that in the n-type semiconductor ML-MoS$_2$ flakes, the concentration of the majority carrier, electron, would be enhanced in the P$_\uparrow$ state locally around the contact area of the AFM tip, while in the P$_\downarrow$ state, electrons would be depleted.

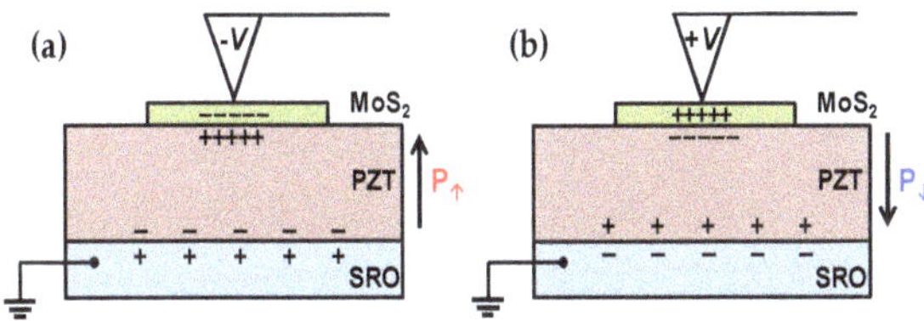

Figure 2. Schematic of the poling process on AFM. (**a**) Up-polarized (P$_\uparrow$) and (**b**) down-polarized (P$_\downarrow$) states are achieved when a PZT thin films are applied by negative and positive V_P above the threshold voltage for polarization reversal using a conductive AFM tip, respectively. During the poling process, a metallic SRO layer is grounded electrically.

Figure 3 summarizes how the PL emission of a ML-MoS$_2$ flake on a PZT thin film can be controlled ferroelectrically by the poling process. The non-volatile PL feature given by the poling process was observed to not be degraded that much up to several days in the air. Figure 3a,d are optical microscopy and AFM topography images taken from the ML-MoS$_2$ flake transferred on the PZT surface. The representative height profile embedded in Figure 3d verifies the thickness of ML-MoS$_2$ while the surface roughness reflects the inherent domain structure of the PZT thin film [8]. Figure 3b,c include the PFM images of the ML-MoS$_2$ flake on the PZT film, in which half of its area is in the P$_\uparrow$ state and the rest is in the opposite P$_\downarrow$ state. From the PL peak area and position maps in Figure 3e–j, the PL intensity of ML-MoS$_2$ is stronger by more than 200% and the PL peak position is more blue-shifted in the P$_\uparrow$ state than in the P$_\downarrow$. The difference in the PL characteristics can be more clearly observed from the PL spectra measured from the spots of the ML-MoS$_2$ flake in the P$_\uparrow$ and P$_\downarrow$ regimes, selectively as shown in Figure 3k. By deconvoluting the PL peaks into two Lorentzian peaks located at ~1.88 eV and ~1.84 eV corresponding to the transitions of X and X$^-$, respectively, the relative contributions of the two different species can be analyzed for each case. While the PL peak is contributed to from both emissions in the P$_\uparrow$ state, the emission of X is suppressed almost completely in the P$_\downarrow$ state. These results show that the control of carrier concentration via FE gating allows PL emissions to be modulated reversibly via exciton-trion transition [8,24]. The same trend was also observed in the ML-MoS$_2$/BFO heterostructures (see Supplementary Figure S2). Further, the microscale PL modulation has been demonstrated more clearly with a stripe poling pattern of a domain-engineered BFO surface (see Supplementary Figure S3). It is worthy to note that, considering the nanoscale size of the AFM tip, the PL modulation can be confined even down to the nanoscale by this approach, as the electrical properties were tuned on ML-MoS$_2$/PZT at the nanoscale by A. Lipatov and his colleagues [27]. However, the trend of PL modulation observed here is opposite to that predicted from the schematics of the poling process in Figure 2 and also that of the previously reported ML-TMD/FE heterostructures whose ML-TMD parts were prepared by mechanical exfoliation from single crystals [8,24]. This inconsistency will be discussed in detail later.

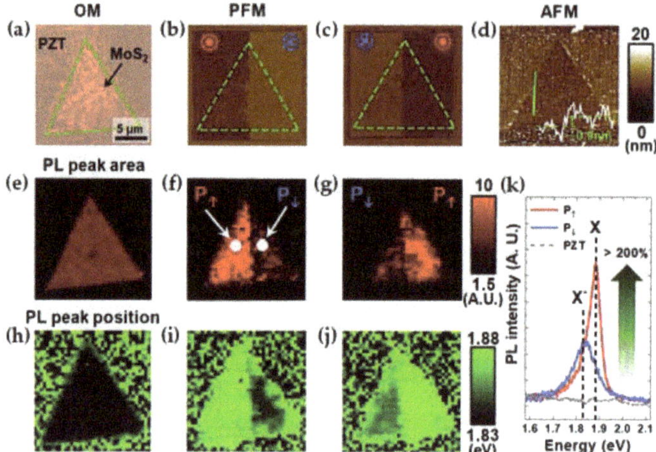

Figure 3. Poling effects on the ML-MoS$_2$/PZT heterostructure. (**a**) Optical microscopy (OM) image of a ML-MoS$_2$ flake with a scale bar which works for all the other images. PFM images of (**b**) the ML-MoS$_2$ flake on the PZT thin film whose left and right half areas are polarized in the P$_\uparrow$ and P$_\downarrow$ states by the poling process with the V_P of −12 V and +12 V, respectively, and (**c**) vice versa. (**d**) Topography AFM image of the ML-MoS$_2$ flake simultaneously obtained with the PFM image including the representative height profile for the corresponding green line, verifying the thickness of ML-MoS$_2$ flake. PL peak area maps of the identical ML-MoS$_2$ flake in (**e**–**g**) were scanned before poling and after the poling processes of (**b**,**c**), in order. (**h**–**j**) PL peak position maps displayed in the same sequence as in (**e**–**g**). (**k**) PL spectra measured from the spots of P$_\uparrow$ and P$_\downarrow$ regions marked in (**f**) along with that of the bare PZT as a reference. X and X$^-$ denote the emissions of neutral and negatively-charged excitons, respectively.

The same trend of polarization-dependent PL has also been observed by electrostatic gating by which the control of polarization of PZT thin films can be performed more efficiently and also more safely than the poling process with the direct contact of AFM tips. As shown in Figure 4a, the metal electrode was fabricated on a ML-MoS$_2$ flake. Figure 4b shows schematically how a PZT thin film can be up-polarized by back-gating with a positive gate voltage (V_G) from a bottom electrode of a SRO thin layer. However, the ML-MoS$_2$ flake seems to be damaged partly through the fabrication process; from the PL peak area maps in Figure 4c–f scanned after the V_G is removed, it can be seen clearly that the PL intensity can be modulated reversibly in a nonvolatile way between the P$_\uparrow$ and P$_\downarrow$ states. The PL modulation is observed to be intensified particularly near the bump inside the dashed circle in Figure 4a where the electric field is focused strongly. Consistently with the results of the first AFM experiment, the PL intensity was observed to be higher in the P$_\uparrow$ state than in the opposite P$_\downarrow$ state. Also, as can be checked in Figure 4g–j, as the PL intensity increases with strengthening up-polarization, the PL peak is blue-shifted slightly, which is probably due to the enhancement of emissions by X.

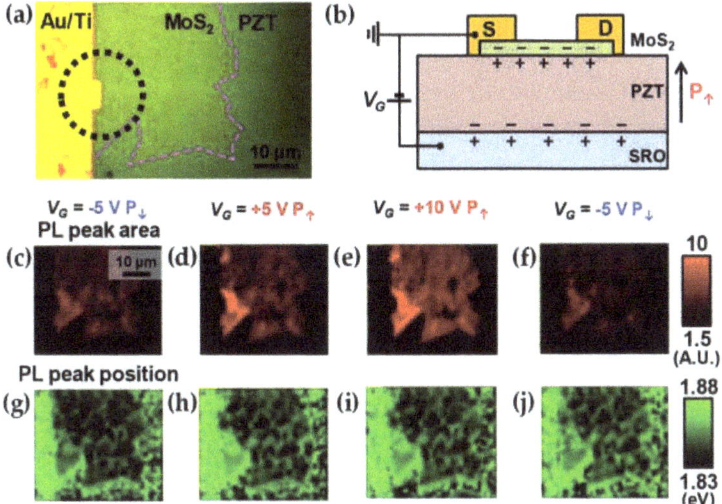

Figure 4. Electrical back-gated experiments on the ML-MoS$_2$/PZT heterostructures. (**a**) Optical microscopy image of a ML-MoS$_2$ sheet with a metal electrode. (**b**) Schematics of back-gating experiment. In this experiment, no voltage is applied between the source and drain. (**c–f**) Set of images of the PL peak area measured after back gating with V_G of −5 V, +5 V, +10 V, and again −5 V in order. (**g–j**) The set of images of PL peak position measured in a series in the same order of (**c–f**).

3.3. Electrical Transport Characterization

To understand the underlying mechanisms of the PL modulation induced by FE polarization deeply, the combined experiments of optical and electrical measurements are also carried out on a FET device based on two ML-MoS$_2$ flakes. The optical microscopy and AFM topography images of the device are shown in Figure 5a,b, respectively. After each poling process, both PL maps and FET characteristic curves were acquired. Figure 5c,d shows that the PL of the ML-MoS$_2$ flakes embedded in the device can also be modulated by the poling process properly. To achieve the P$_\uparrow$ and P$_\downarrow$ states, the area including the ML-MoS$_2$ region was scanned, applying V_P of −12 V and +12 V with a AFM tip, respectively. Figure 5e shows schematics of the device with the circuit for the FET measurements. As shown in the drain current (I_D) vs. V_G plots in Figure 5f, in the both the P$_\uparrow$ and P$_\downarrow$ states, the typical n-type FET characteristics are observed with a large hysteresis probably due to the environmental effects [38]. Also, it can be seen that the very high dielectric constant of the PZT enables low-power FET operation with very small V_G [8,39]. The conduction level is higher in the P$_\downarrow$ state than the P$_\uparrow$ state by almost an order of magnitude implying that the electron concentration is also higher in the P$_\downarrow$ state, which is consistent with the analysis with the PL spectra. In addition, the variation of leakage current (I_G) with the polarization direction is likely to be driven by the polarization-induced band structure change across the ML-MoS$_2$/PZT junction [26].

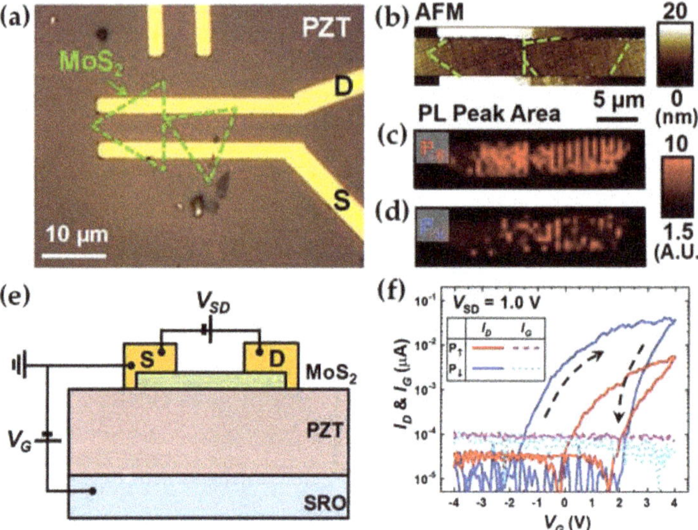

Figure 5. Electrical transport characterization of the ML-MoS$_2$/PZT heterostructures. (**a**) Optical microscopy image and (**b**) topography AFM image of a device with a channel based on two ML-MoS$_2$ sheets and metal electrodes on a PZT thin film. PL peak area maps of (**c,d**) were scanned after the poling process with V_P of −12 V and +12 V, respectively. (**e**) Device schematics for FET measurements with the circuit. (**f**) FET characteristic curves of drain current (I_D) vs. gate voltage (V_G) measured at the source-drain voltage (V_{SD}) of 1.0 V in the P$_\uparrow$ and P$_\downarrow$ states, respectively. The leakage current (I_G) vs. V_G are also plotted as dashed lines for both states. Note that the absolute values were taken for I_D and I_G.

Now, based on all these results, it will be discussed why the trends of FE effects observed in this work are contrast to those of conventional FeFETs [8,19,27]. As in the first scenario, positively-charged exciton X$^+$, which is expected to have similar PL characteristics to those of negatively-charged exciton X$^-$, may be excited dominantly in the P$_\downarrow$ state as inversion occurs in the ML-MoS$_2$ flake [5]. While A. Lipatov et al. argued that the majority carriers of ML-MoS$_2$ on PZT or BTO would be holes, in this case, electrons should be majority carriers from the n-type conduction of the FET devices as shown in Figure 5f [27]. The n-type conduction is even stronger in the P$_\downarrow$ state than that in the P$_\uparrow$ state, ruling out the possibility of the dominance of X$^+$. Moreover, from the FET measurements, only clockwise FET hysteresis loops were observed. Typical counterclockwise operation for FeFETs caused by spontaneous polarization and abrupt charging/discharging upon polarization reversal were not observed even from the FET characterization in the wider V_G range well above the threshold voltage of PZT thin films of ~3–4 V (see Supplementary Figure S4) [8,19].

Along the van der Waals interfaces between the CVD-grown ML-MoS$_2$ layers and the FE surface, the interfacial traps and contaminants can be formed during the wet-transfer process for fabricating heterostructures and may mitigate the interaction across the 2D/FE interface [14]. The significant hysteresis of FET characteristic loops indicates a strong influence of environmental factors such as molecular adsorption, humidity, interfacial traps, and so on [14,38,40]. Further, maybe due to the complicated interplay among inherent defects of the CVD-grown ML-MoS$_2$ and the interfacial traps and contaminants, abnormal defect-related interactions coupled with FE polarization may be activated. For example, the ionic species of the interfacial trapped water layer are likely to be injected into or removed from the ML-MoS$_2$ surface through the reversible electrochemical process induced by FE polarization [21,41]. It is known that O$_2$ or water molecules take away electrons from ML-MoS$_2$ more actively around point defects such as S vacancies [5,29,31]. Accordingly, the ionic species seem to charge or discharge the ML-MoS$_2$ interacting with the point defects differently depending on the polarization state. Moreover, the ambient molecules such as O_2 and water vapor may be involved in the modulation of carrier concentrations in ML-MoS$_2$. From the results of optical and electrical characterization, the charge transfer from ML-MoS$_2$ to trapped ionic species is likely to be more active in the P$_\uparrow$ state leading to a decrease in the carrier concentration and eventually, the enhancement of emissions of X. To verify the trap effects as well as facilitate the FE effects fully, trap-free van der Waals interfaces should be achieved possibly by assembling the heterostructures using the dry-transfer method in the inert gas environment [14]. Therefore, the ML MoS$_2$/FE heterostructures may also have potential to be applicable to ionic sensors besides optoelectronic devices.

3.4. Atomically-Thin Electric Field Barrier

Lastly, ML-MoS$_2$ flakes have been evaluated for whether they work well as electric field barriers using 2D/FE hybrid platforms under the poling process in air or not. Figure 6 shows that ML-MoS$_2$ prevents electric field penetration into a BFO thin film quite well considering its sub-nanometer thickness. While ML-MoS$_2$ is quite conductive with the Fermi energy level located very closely to the conduction band edge in a vacuum, it becomes less conductive in air along with the shift of the Fermi energy toward the midgap energy level via interacting with ambient gas such as O_2 and water vapor [40]. Figure 6a shows that the PFM image scanned from just after the ML-MoS$_2$ flake was transferred on the unpoled region of the BFO thin film which shows the irregular polarization pattern of the pristine BFO thin film. Since the ML-MoS$_2$ sheet is ultrathin and almost transparent to the piezoelectric response, the pattern can be seen through the ML-MoS$_2$ flake clearly. Up to V_P of +8 V, the region of the BFO thin film underneath the ML-MoS$_2$ flake was not poled well, indicating that the ML-MoS$_2$ can block the electric field significantly. The ML-MoS$_2$ regime was down-polarized finally after poling with V_P of +10 V. The electric field was observed to be shielded during the poling process to achieve the opposite up-polarized states on the same heterostructure. Moreover, also in the case of the ML-MoS$_2$/PZT system, similar electric field screening effects were observed (see Supplementary Figure S5). Now, it is clearly understood why PL modulation can be achieved upon polarization reversal only when the application of V_P (or V_G) is well above the threshold voltage of ~3–4 V, which is determined by the coercive field of the FE films.

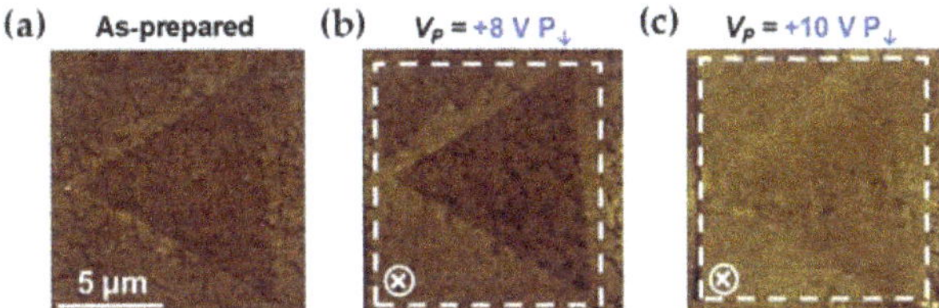

Figure 6. The ultrathin electric field shield. PFM images of a ML-MoS$_2$ flake on a BFO thin film scanned (**a**) as-prepared before poling and after poling with the V_P of (**b**) +8 V and (**c**) +10 V, respectively. Up to V_P of +8 V, the area of the BFO thin film beneath the ML-MoS$_2$ flake was not poled well, indicating that a ML-MoS$_2$ flake prevents field penetration into the BFO thin film.

4. Conclusions

In this research, the alteration of nonvolatile PL over 200% has been achieved upon polarization reversal along with large modulation in electrical properties in the use of lateral ML-MoS$_2$/FE heterostructures. Based on comprehensive experiments and analyses, the spontaneous polarization of the FE thin films seems to affect the optoelectronic behaviors of ML-MoS$_2$ indirectly via reversible electrochemical processes among interfacial traps, air molecules, and structural imperfections. In addition, in the same geometry, it was also shown that MoS$_2$ can shield the electric field effectively even with sub-nanometer thickness in ambient conditions.

Supplementary Materials: The following are available online at http://www.mdpi.com/2079-4991/9/11/1620/s1, Figure S1: Optical characterization of the ML-MoS$_2$/BFO heterostructure. Figure S2: Poling effects on the ML-MoS$_2$/BFO heterostructure. Figure S3: Microscale PL modulation of the ML-MoS$_2$ driven by the domain-engineered BFO thin film. Figure S4: Field-effect-transistor characteristics of the ML-MoS$_2$/PZT heterostructure device. Figure S5: Electric field screening effects of the ML-MoS$_2$ on PZT.

Author Contributions: C.K. conceived the research and designed the experiments. C.K. carried out all optical and electrical measurements as well as data analyses. Most of important processes for materials preparation and device fabrication were also performed by C.K.

Funding: This research was funded by Sookmyung Women's University Research Grants (1-1703-2003) and Basic Science Research Program through the National Research Foundation of Korea (NRF) funded by the Ministry of Education (NRF-2017R1D1A1B03036223). The e-beam evaporation process in this research was partly supported by Nano·Material Technology Development Program through the National Research Foundation of Korea (NRF) funded by the Ministry of Science, ICT and Future Planning. (2009-0082580).

Acknowledgments: The author thanks Junqiao Wu for useful discussion and Yabin Chen, Yeonbae Lee, and Deyang Chen for technical assistance.

Conflicts of Interest: The author declares no conflict of interest.

References

1. Mak, K.F.; Lee, C.; Hone, J.; Shan, J.; Heinz, T.F. Atomically thin MoS$_2$: A new direct-gap semiconductor. *Phys. Rev. Lett.* **2010**, *105*, 136805. [CrossRef]

2. Mak, K.F.; He, K.; Lee, C.; Lee, G.H.; Hone, J.; Heinz, T.F.; Shan, J. Tightly bound trions in monolayer MoS$_2$. *Nat. Mater.* **2013**, *12*, 207–211. [CrossRef] [PubMed]

3. Chernikov, A.; Berkelbach, T.C.; Hill, H.M.; Rigosi, A.; Li, Y.; Aslan, O.B.; Reichman, D.R.; Hybertsen, M.S.; Heinz, T.F. Exciton binding energy and nonhydrogenic Rydberg series in monolayer WS$_2$. *Phys. Rev. Lett.* **2014**, *113*, 076802. [CrossRef] [PubMed]

4. Zeng, H.; Dai, J.; Yao, W.; Xiao, D.; Cui, X. Valley polarization in MoS$_2$ monolayers by optical pumping. *Nat. Nanotechnol.* **2012**, *7*, 490–493. [CrossRef] [PubMed]

5. Tongay, S.; Zhou, J.; Ataca, C.; Lo, K.; Matthews, T.S.; Li, J.; Grossman, J.C.; Wu, J. Thermally driven crossover from indirect toward direct bandgap in 2D semiconductors: MoSe$_2$ versus MoS$_2$. *Nano Lett.* **2012**, *12*, 5576–5580. [CrossRef] [PubMed]

6. Ross, J.S.; Klement, P.; Jones, A.M.; Ghimire, N.J.; Yan, J.; Mandrus, D.; Taniguchi, T.; Watanabe, K.; Kitamura, K.; Yao, W. Electrically tunable excitonic light-emitting diodes based on monolayer WSe$_2$ p–n junctions. *Nat. Nanotechnol.* **2014**, *9*, 268–272. [CrossRef] [PubMed]

7. Wachter, S.; Polyushkin, D.K.; Bethge, O.; Mueller, T. A microprocessor based on a two-dimensional semiconductor. *Nat. Commun.* **2017**, *8*, 14948. [CrossRef] [PubMed]

8. Ko, C.; Lee, Y.; Chen, Y.; Suh, J.; Fu, D.; Suslu, A.; Lee, S.; Clarkson, J.D.; Choe, H.S.; Tongay, S. Ferroelectrically gated atomically thin transition-metal dichalcogenides as nonvolatile memory. *Adv. Mater.* **2016**, *28*, 2923–2930. [CrossRef]

9. Park, M.; Park, Y.J.; Chen, X.; Park, Y.; Kim, M.; Ahn, J. MoS$_2$-based tactile sensor for electronic skin applications. *Adv. Mater.* **2016**, *28*, 2556–2562. [CrossRef]

10. Ge, R.; Wu, X.; Kim, M.; Shi, J.; Sonde, S.; Tao, L.; Zhang, Y.; Lee, J.C.; Akinwande, D. Atomristor: Nonvolatile resistance switching in atomic sheets of transition metal dichalcogenides. *Nano Lett.* **2017**, *18*, 434–441. [CrossRef]

11. Gao, L. Flexible device applications of 2D semiconductors. *Small* **2017**, *13*, 1603994. [CrossRef] [PubMed]

12. Yuan, Z.; Hou, J.; Liu, K. Interfacing 2D semiconductors with functional oxides: Fundamentals, properties, and applications. *Crystals* **2017**, *7*, 265. [CrossRef]

13. Jariwala, D.; Marks, T.J.; Hersam, M.C. Mixed-dimensional van der Waals heterostructures. *Nat. Mater.* **2017**, *16*, 170–181. [CrossRef] [PubMed]

14. Frisenda, R.; Navarro-Moratalla, E.; Gant, P.; De Lara, D.P.; Jarillo-Herrero, P.; Gorbachev, R.V.; Castellanos-Gomez, A. Recent progress in the assembly of nanodevices and van der Waals heterostructures by deterministic placement of 2D materials. *Chem. Soc. Rev.* **2018**, *47*, 53–68. [CrossRef] [PubMed]

15. Li, S.; Wakabayashi, K.; Xu, Y.; Nakaharai, S.; Komatsu, K.; Li, W.; Lin, Y.; Aparecido-Ferreira, A.; Tsukagoshi, K. Thickness-dependent interfacial coulomb scattering in atomically thin field-effect transistors. *Nano Lett.* **2013**, *13*, 3546–3552. [CrossRef] [PubMed]

16. Howell, S.L.; Jariwala, D.; Wu, C.; Chen, K.; Sangwan, V.K.; Kang, J.; Marks, T.J.; Hersam, M.C.; Lauhon, L.J. Investigation of band-offsets at monolayer–multilayer MoS_2 junctions by scanning photocurrent microscopy. *Nano Lett.* **2015**, *15*, 2278–2284. [CrossRef] [PubMed]

17. Hou, J.; Wang, X.; Fu, D.; Ko, C.; Chen, Y.; Sun, Y.; Lee, S.; Wang, K.X.; Dong, K.; Sun, Y. Modulating photoluminescence of monolayer molybdenum disulfide by metal–insulator phase transition in active substrates. *Small* **2016**, *12*, 3976–3984. [CrossRef]

18. Chen, J.; Odenthal, P.M.; Swartz, A.G.; Floyd, G.C.; Wen, H.; Luo, K.Y.; Kawakami, R.K. Control of Schottky barriers in single layer MoS_2 transistors with ferromagnetic contacts. *Nano Lett.* **2013**, *13*, 3106–3110. [CrossRef]

19. Mathews, S.; Ramesh, R.; Venkatesan, T.; Benedetto, J. Ferroelectric Field Effect Transistor Based on Epitaxial Perovskite Heterostructures. *Science* **1997**, *276*, 238–240. [CrossRef]

20. Bruchhaus, R.; Pitzer, D.; Schreiter, M.; Wersing, W. Optimized PZT thin films for pyroelectric IR detector arrays. *J. Electroceram.* **1999**, *3*, 151–162. [CrossRef]

21. Kakekhani, A.; Ismail-Beigi, S. Ferroelectric oxide surface chemistry: Water splitting via pyroelectricity. *J. Mater. Chem. A* **2016**, *4*, 5235–5246. [CrossRef]

22. Khan, M.A.; Nadeem, M.A.; Idriss, H. Ferroelectric polarization effect on surface chemistry and photo-catalytic activity: A review. *Surf. Sci. Rep.* **2016**, *71*, 1–31. [CrossRef]

23. Arimoto, Y.; Ishiwara, H. Current status of ferroelectric random-access memory. *MRS Bull.* **2004**, *29*, 823–828. [CrossRef]

24. Wen, B.; Zhu, Y.; Yudistira, D.; Boes, A.; Zhang, L.; Yildirim, T.; Liu, B.; Yan, H.; Sun, X.; Zhou, Y. Ferroelectric Driven Exciton and Trion Modulation in Monolayer Molybdenum and Tungsten Diselenides. *ACS Nano* **2019**, *13*, 5335–5343. [CrossRef] [PubMed]

25. Si, M.; Liao, P.; Qiu, G.; Duan, Y.; Ye, P.D. Ferroelectric field-effect transistors based on MoS_2 and CuInP2S6 two-dimensional van der Waals heterostructure. *ACS Nano* **2018**, *12*, 6700–6705. [CrossRef]

26. Li, T.; Sharma, P.; Lipatov, A.; Lee, H.; Lee, J.; Zhuravlev, M.Y.; Paudel, T.R.; Genenko, Y.A.; Eom, C.; Tsymbal, E.Y. Polarization-mediated modulation of electronic and transport properties of hybrid MoS_2–$BaTiO_3$–$SrRuO_3$ tunnel junctions. *Nano Lett.* **2017**, *17*, 922–927. [CrossRef]

27. Lipatov, A.; Li, T.; Vorobeva, N.S.; Sinitskii, A.; Gruverman, A. Nanodomain Engineering for Programmable Ferroelectric Devices. *Nano Lett.* **2019**, *19*, 3194–3198. [CrossRef]

28. Bao, W.; Borys, N.J.; Ko, C.; Suh, J.; Fan, W.; Thron, A.; Zhang, Y.; Buyanin, A.; Zhang, J.; Cabrini, S. Visualizing nanoscale excitonic relaxation properties of disordered edges and grain boundaries in monolayer molybdenum disulfide. *Nat. Commun.* **2015**, *6*, 7993. [CrossRef]

29. Lee, Y.; Park, S.; Kim, H.; Han, G.H.; Lee, Y.H.; Kim, J. Characterization of the structural defects in CVD-grown monolayered MoS_2 using near-field photoluminescence imaging. *Nanoscale* **2015**, *7*, 11909–11914. [CrossRef]

30. Nan, H.; Wang, Z.; Wang, W.; Liang, Z.; Lu, Y.; Chen, Q.; He, D.; Tan, P.; Miao, F.; Wang, X. Strong photoluminescence enhancement of MoS_2 through defect engineering and oxygen bonding. *ACS Nano* **2014**, *8*, 5738–5745. [CrossRef]

31. Tongay, S.; Suh, J.; Ataca, C.; Fan, W.; Luce, A.; Kang, J.S.; Liu, J.; Ko, C.; Raghunathanan, R.; Zhou, J. Defects activated photoluminescence in two-dimensional semiconductors: Interplay between bound, charged, and free excitons. *Sci. Rep.* **2013**, *3*, 2657. [CrossRef] [PubMed]

32. Kang, K.; Xie, S.; Huang, L.; Han, Y.; Huang, P.Y.; Mak, K.F.; Kim, C.; Muller, D.; Park, J. High-mobility three-atom-thick semiconducting films with wafer-scale homogeneity. *Nature* **2015**, *520*, 656–660. [CrossRef] [PubMed]

33. Ramesh, R.; Aggarwal, S.; Auciello, O. Science and technology of ferroelectric films and heterostructures for non-volatile ferroelectric memories. *Mater. Sci. Eng. R* **2001**, *32*, 191–236. [CrossRef]

34. Lin, C.; Shih, W.; Chang, I.Y.; Juan, P.; Lee, J.Y. Metal-ferroelectric ($BiFeO_3$)-insulator (Y_2O_3)-semiconductor capacitors and field effect transistors for nonvolatile memory applications. *Appl. Phys. Lett.* **2009**, *94*, 142905. [CrossRef]

35. Tian, G.; Zhao, L.; Lu, Z.; Yao, J.; Fan, H.; Fan, Z.; Li, Z.; Li, P.; Chen, D.; Zhang, X. Fabrication of high-density $BiFeO_3$ nanodot and anti-nanodot arrays by anodic alumina template-assisted ion beam etching. *Nanotechnology* **2016**, *27*, 485302. [CrossRef]

36. Yang, L.; Majumdar, K.; Liu, H.; Du, Y.; Wu, H.; Hatzistergos, M.; Hung, P.; Tieckelmann, R.; Tsai, W.; Hobbs, C. Chloride molecular doping technique on 2D materials: WS_2 and MoS_2. *Nano Lett.* **2014**, *14*, 6275–6280. [CrossRef]

37. Suh, J.; Park, T.; Lin, D.; Fu, D.; Park, J.; Jung, H.J.; Chen, Y.; Ko, C.; Jang, C.; Sun, Y. Doping against the native propensity of MoS_2: Degenerate hole doping by cation substitution. *Nano Lett.* **2014**, *14*, 6976–6982. [CrossRef]

38. Late, D.J.; Liu, B.; Matte, H.R.; Dravid, V.P.; Rao, C. Hysteresis in single-layer MoS_2 field effect transistors. *ACS Nano* **2012**, *6*, 5635–5641. [CrossRef]

39. Akekhani, A.; Ismail-Beigi, S. Polarization-driven catalysis via ferroelectric oxide surfaces. *Phys. Chem. Chem. Phys.* **2016**, *18*, 19676–19695. [CrossRef]

40. Lee, S.Y.; Kim, U.J.; Chung, J.; Nam, H.; Jeong, H.Y.; Han, G.H.; Kim, H.; Oh, H.M.; Lee, H.; Kim, H. Large work function modulation of monolayer MoS_2 by ambient gases. *ACS Nano* **2016**, *10*, 6100–6107. [CrossRef]

41. Kalinin, S.V.; Jesse, S.; Tselev, A.; Baddorf, A.P.; Balke, N. The role of electrochemical phenomena in scanning probe microscopy of ferroelectric thin films. *ACS Nano* **2011**, *5*, 5683–5691. [CrossRef]

Article

Characteristics of p-Type Conduction in P-Doped MoS_2 by Phosphorous Pentoxide during Chemical Vapor Deposition

Jae Sang Lee, Chang-Soo Park, Tae Young Kim, Yoon Sok Kim and Eun Kyu Kim *

Department of Physics and Research Institute of Natural Sciences, Hanyang University, Seoul 04763, Korea
* Correspondence: ek-kim@hanyang.ac.kr

Received: 19 August 2019; Accepted: 5 September 2019; Published: 7 September 2019

Abstract: We demonstrated p-type conduction in MoS_2 grown with phosphorous pentoxide via chemical vapor deposition (CVD). Monolayer MoS_2 with a triangular shape and 15-μm grains was confirmed by atomic force microscopy. The difference between the Raman signals of the A_{1g} and E^1_{2g} modes for both the pristine and P-doped samples was 19.4 cm^{-1}. In the X-ray photoelectron spectroscopy results, the main core level peaks of P-doped MoS_2 downshifted by about 0.5 eV to a lower binding energy compared to the pristine material. Field-effect transistors (FETs) fabricated with the P-doped monolayer MoS_2 showed p-type conduction with a field-effect mobility of 0.023 cm^2/V·s and an on/off current ratio of 10^3, while FETs with the pristine MoS_2 showed n-type behavior with a field-effect mobility of 29.7 cm^2/V·s and an on/off current ratio of 10^5. The carriers in the FET channel were identified as holes with a concentration of 1.01×10^{11} cm^{-2} in P-doped MoS_2, while the pristine material had an electron concentration of 6.47×10^{11} cm^{-2}.

Keywords: chemical vapor deposition; P_2O_5; p-type conduction; P-doped MoS_2

---

1. Introduction

Recently, various studies have analyzed two-dimensional (2D) materials, such as graphene, MoS_2, and WSe_2, because of their critical properties and abundant potential for use in optical and electrical applications [1–3]. Graphene has a zero band gap structure, but has not been able to replace semiconductor-based devices [4,5]. Additionally, layered transition metal dichalcogenides (TMDs) such as MoS_2 and WSe_2 have received enormous attention as promising materials and layer structures, in which transition metals are sandwiched between two chalcogen atom layers by a covalent force. Moreover, there are Van der Waals (VdW) forces interacting in individual layers, which make exfoliation easily. Interestingly, these materials have a unique property; their band gap structure varies depending on the thickness. In the case of MoS_2, the band gap of a monolayer has a direct band gap of 1.8 eV, while a few layers of MoS_2 and bulk MoS_2 have an indirect band gap structure with a band gap of about 1.2 eV [6].

The chemical vapor deposition (CVD) method has several advantages compared to other methods, such as mechanical and liquid exfoliation methods [7,8]. The disadvantages of the Scotch tape-based mechanical method are its difficulty in controlling the flake thickness, size, and uniformity, which makes it inappropriate for large-scale applications. The liquid method still needs to be developed for applications, while the CVD method has been used to prepare ultrathin monolayers or few-layer MoS_2 films over large areas [9]. Transistors have been fabricated via CVD growth of monolayer MoS_2. These have been reported to exhibit good properties, including a high on/off current ratio and high mobility [10]. To realize detailed applications, this method needs to be able to produce a junction composed of n- and p-type materials. Although there have been many challenges to p-type doping of MoS_2 using niobium (Nb) or phosphorous (P) atoms [11,12], and it remains difficult to successfully dope ultrathin MoS_2. According to a previous report, P atoms seem to be the most suitable acceptors among group V elements [13].

In this paper, we report on the CVD growth and characteristics of monolayer MoS_2 with and without the addition of phosphorous pentoxide (P_2O_5) powder. The thickness and grain size of the MoS_2 layer were measured using non-contact-mode atomic force microscopy (AFM) and Raman spectroscopy. To confirm the electrical characteristics of MoS_2, back-gated field-effect transistors (FETs) were fabricated. The p-type conduction from monolayer MoS_2 grown with P_2O_5 powder was confirmed and compared to pristine MoS_2 with n-type behavior.

2. Experimental

To synthesize an MoS_2 layer by the CVD method, molybdenum trioxide (MoO_3, CERAC Inc, Milwaukee, WI, USA) powder with 99.999% purity as a precursor material and sulfur (iTASCO Inc, Seoul, Korea) powder of 99.999% purity as a reactant material were used. For p-type doping of MoS_2 in this experiment, 98.99% purity P_2O_5 (SIGMA-ALDRICH, St. Louis, MO, USA) powder was added as a dopant material. SiO_2/Si substrates (2×2 cm^2) with a SiO_2 thickness of 270 nm and three alumina boats were used. The alumina boats were filled with 10 mg of MoO_3 powder, 300 mg of S powder, and 1 mg of P_2O_5 powder, respectively. During CVD growth of MoS_2, the furnace was heated to 750 °C with a heating rate of 30 °C/min under argon gas flowing at 100 sccm. The role of argon gas was to transport S and P_2O_5 when they were vaporized. During the growth of MoS_2, the gas flow and furnace temperature were kept constant for 30 min, and then the furnace was quickly cooled down to room temperature.

The MoS_2 thickness and grain size were analyzed by using non-contact-mode atomic force microscopy (AFM) (XE-100, Park's Systems, Seoul, Korea) and optical microscopy. X-ray photoelectron spectroscopy (XPS) (K-Alpha+, Thermo Fisher Scientific, Waltham, MA, USA) under ~4×10^{-10} Torr and Raman spectroscopy (NRS-3100, JASCO, Tokyo, Japan) with a $\lambda = 532$ nm laser were measured at room temperature to identify the doping characteristics. To confirm the electrical characteristics of doped monolayer MoS_2, back-gated FETs were fabricated. In this process, photolithography was used for patterning source and drain electrodes of Ni/Au (5 nm/50 nm) metals.

3. Results

Figure 1a shows a simplified schematic diagram for the synthesis of MoS_2, with and without phosphorus doping, using the CVD system. Here, MoO_3 powder was placed in the middle of the furnace, slightly away from the S and P_2O_5 powders. The P_2O_5 powder for P doping was located about 7.5 cm from the MoO_3 powder. The ratio of S to Mo atoms is an important point for growing monolayer MoS_2 flakes. We used a face-down substrate approach, where the SiO_2 substrate is positioned vertically facing the MoO_3-containing alumina boat. Unlike previous doping studies that used a two-furnace system [14–16], this method used in situ doping with a one-furnace CVD system. The CVD process for MoS_2 growth can be divided into two steps: Nucleation and growth. Figure 1b shows the temperature profile of the reaction furnace and pressure variation in the quartz tube with Ar gas flow, respectively, as a function of time. In this figure, S atoms are introduced at 650 °C which is 100 °C lower than the growth temperature (750 °C). When S atoms are introduced, the growth of MoS_2 starts and then monolayer MoS_2 flakes appear [17].

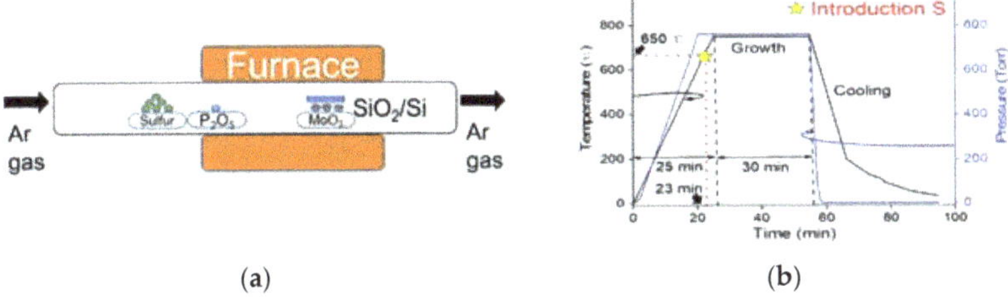

(a)   (b)

Figure 1. (**a**) Schematic diagram of the chemical vapor deposition (CVD) process for the monolayer MoS$_2$ synthesis and in situ P doping with P$_2$O$_5$ powder. (**b**) Temperature profile of the reaction furnace and pressure in the quartz tube as a function of the processing time.

Figure 2a shows an optical microscopy image of P-doped MoS$_2$ grown via CVD. Here, the MoS$_2$ layers grown on the SiO$_2$/Si substrate under a sufficient S atmosphere were observed to have a triangular shape [17,18]; this is the same shape as pristine MoS$_2$. The grain size of doped MoS$_2$ on the SiO$_2$/Si substrate was about 15 μm. To confirm the formation of a monolayer of P-doped MoS$_2$, Raman spectroscopy and AFM measurements were performed, as shown in Figure 2b,c, respectively. The thickness of an MoS$_2$ flake measured by AFM was about 0.6 nm to 0.9 nm; this layer thickness is the same as a previous result [10]. This measurement value corresponds to the interlayer spacing of a monolayer of S-Mo-S bonding in the MoS$_2$ crystal. Two characteristic Raman peaks, i.e., E$^1_{2g}$ and A$_{1g}$ from in-plane and out-of-plane modes, respectively, were measured by a laser with an excitation wavelength of 532 nm at room temperature, as shown in Figure 2c. The in-plane E$^1_{2g}$ mode presents the vibration of one Mo atom and two S atoms in opposite directions, while the out-of-plane A$_{1g}$ mode vibrates only S atoms in opposite directions (as shown in the inset of Figure 2c). From reported results that describe the dependence of the Raman peaks on the number of layers [19,20], we know that the difference between two Raman peaks depending on the number of MoS$_2$ layers is larger than 20 cm^{-1} for thicknesses above a bilayer (2 L). As shown in Figure 2d, Raman peaks from P-doped MoS$_2$ were located at 384.5 cm^{-1} (E$^1_{2g}$ mode) and 403.9 cm^{-1} (A$_{1g}$ mode). On the other hand, the E$^1_{2g}$ and A$_{1g}$ signals of the pristine monolayer MoS$_2$ were observed at around 384.6 cm^{-1} and 405 cm^{-1}, respectively. The difference between the two Raman modes for P-doped and pristine MoS$_2$ (Figure 2d) appear to be about 19.4 cm^{-1} and 20 cm^{-1}, respectively; these values indicate a single layer of MoS$_2$.

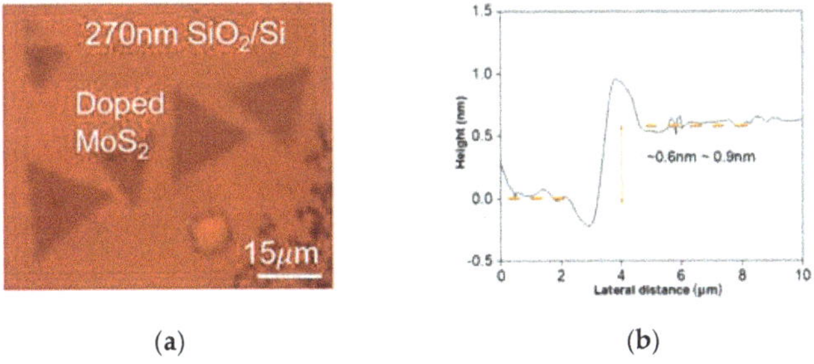

(a)   (b)

Figure 2. *Cont.*

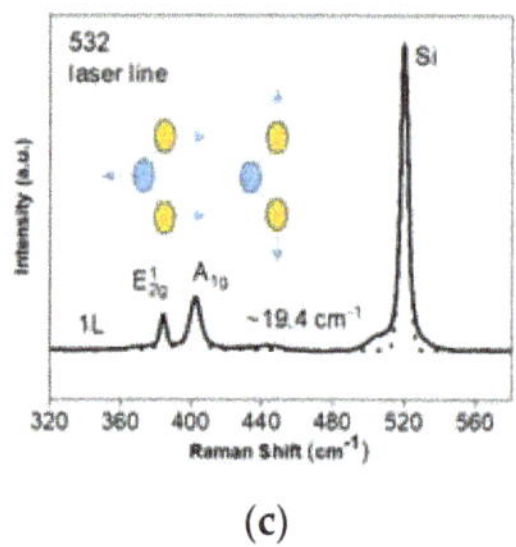

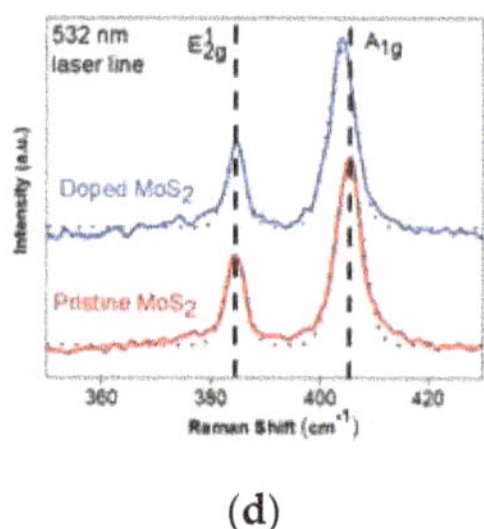

(c)   (d)

Figure 2. (**a**) Optical microscope image, (**b**) AFM height profile, and (**c**) Raman spectroscopy results using a laser with an excitation wavelength of 532 nm for a monolayer of CVD-grown MoS_2 flakes. (**d**) The two Raman modes for the pristine and doped monolayer MoS_2 flakes.

Here, the Raman signal peak of the A_{1g} mode was found to be shifted by about 1.1 cm^{-1}, while the signal peak of the E^1_{2g} mode was almost unchanged. Azcatl et al. reported that a strain induced by dopants can generate contractions of the MoS_2 lattice structure [21]; this phenomenon occurs due to the longer bond length of Mo-S atoms than that of Mo-P atoms. It was also reported that the A_{1g} mode is often more influenced by doping effects than other modes (e.g., the E^1_{2g} mode); this is due to its strong coupling with electrons [22,23]. Therefore, the Raman active signal with A_{1g} has a shift larger than the other Raman active signal because this peak of the Raman mode is quite sensitive to the doping effect. We confirmed that the Raman shifts in Figure 2d agreed with previous results [24]. The full width at half maximum (FHWM) of the E^1_{2g} peak was investigated to characterize the crystalline quality of MoS_2 obtained by the CVD synthesis method. The FWHM result of the CVD-grown monolayer flake is 3.8 cm^{-1}, which is similar to a recently reported value of a CVD-synthesized single-layer flake [18].

The energy peaks appearing in XPS were also analyzed to confirm the doping properties in monolayer MoS_2 crystals. Figure 3a–c show the comparative XPS core level analyses of pristine and doped monolayer MoS_2. In Figure 3a, the P 2p binding energy peak, which clearly appears only at 134.3 eV, is associated with a doped flake feature. It is worth mentioning that the existence of this peak provides apparent evidence that P_2O_5 takes its position before the introduction of S. In addition, the Mo 3d and S 2p core levels indicated that the phenomenon causes a uniform shift of 0.5 eV, from 229.6 eV to 229.1 eV and from 162.4 eV to 161.9 eV, respectively (Figure 3b,c). That is, each peak moved toward a lower binding energy after P-doping, which is very similar to the reported results for Nb-doped MoS_2 [24]. This study reported that the Fermi level (E_F) of pristine MoS_2 is located close to the conduction band (E_c) edge, while an Nb-doped p-type sample has a Fermi level near the valence band edge. The work function and electron affinity of the pristine monolayer MoS_2 is 5.1 eV and 4.28 eV, respectively [25]. The pristine MoS_2 Fermi level is 0.82 eV, which is the E_c–E_F result, and the doping sample Fermi level is 1.32 eV, 0.82 + ΔBE (measured from XPS data). Therefore, it is suggested that doping with P_2O_5 leads to a downshift in the Fermi level of about 0.5 eV, close to the valence band maximum.

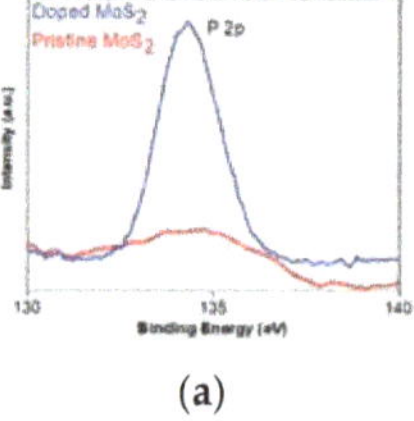

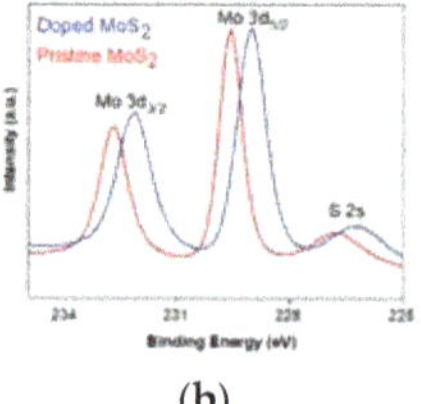

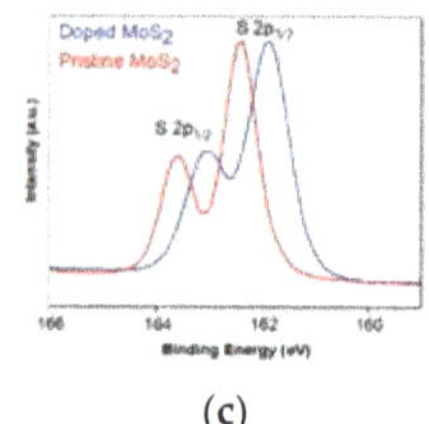

(a)   (b)   (c)

Figure 3. XPS spectra of (**a**) P 2p, (**b**) Mo 3d, and (**c**) S 2p peaks in the pristine and doped MoS_2. These results indicate that the peaks of each core level are downshifted in the doped MoS_2 flake.

Figure 4a,b show the fabricated back-gate FET schematic with a channel length of 3 μm and a channel width of 10 μm, as well as the I_{DS}–V_{DS} curve of an FET based on P-doped monolayer MoS$_2$, respectively.

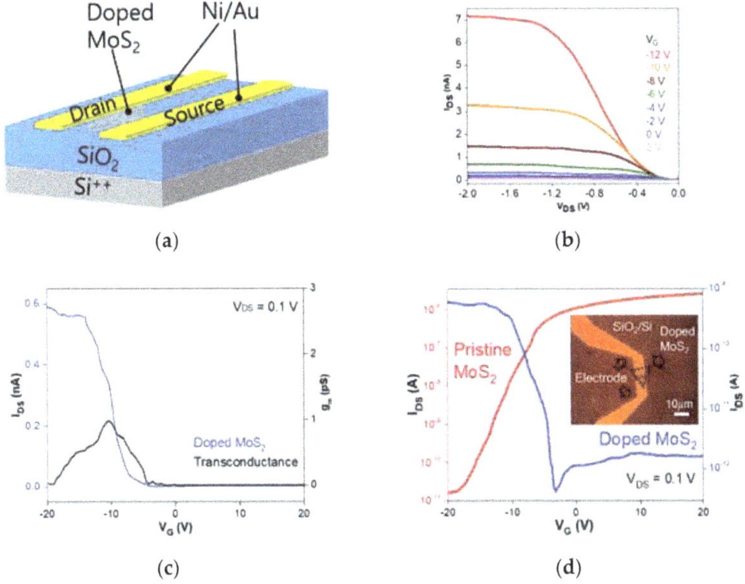

Figure 4. (**a**) Schematic of field-effect transistors (FETs) with a channel length of 3 nm and a channel width of 10 nm. (**b**) I_{DS}–V_{DS} curves of a P-doped monolayer MoS$_2$ FET with different gate voltages. (**c,d**) Linear and log scales of the transfer characteristics as a function of the gate voltage for FETs with pristine and P-doped MoS$_2$ channels, respectively.

Here, Ohmic metals of Ti and Ni were used for n-type pristine and p-type P-doped MoS$_2$ FETs, respectively, to match the metal work functions [26]. Figure 4c shows the transfer characteristics of these devices fabricated on pristine and P-doped monolayer MoS$_2$ flakes. The inset image of Figure 4d is the optical microscopy image of MoS$_2$ FETs which was fabricated on the doped monolayer MoS$_2$. The pristine MoS$_2$ FETs demonstrates n-type conduction with a high on/off current ratio of ~10^5 [27]. The threshold voltage V_T value extracted by the linear extrapolation method was about −8.1 V. In the case of P-doping, the transfer curve indicated p-type conduction with an on/off current ratio of ~10^3, and the V_T was −6.9 V at a drain-source voltage of 0.1 V. The field-effect mobilities of these FETs were calculated by the following relation:

$$\mu = (dI_{DS}/dV_{BG}) \times [L/C_{ox} WV_{DS}], \tag{1}$$

where L and W are the channel length and width, respectively. The back-gate capacitance ($C_{ox} = \varepsilon_0\varepsilon_r/d$) was ~$1.28 \times 10^{-8}$ F/cm^2, where ε_{0X} is the dielectric constant and d is the thickness of silicon oxide. Using the transconductance value obtained by the relation of $g_m = dI_{DS}/dV_{BG}$, the field-effect mobilities were determined to be about 29.7 cm^2/V·s and 0.023 cm^2/V·s for the pristine and P-doped MoS$_2$ FETs, respectively. The carrier concentration in the FET channel could also be estimated by using the following relation:

$$n = C_{ox} (V_{BG} - V_T)/e, \tag{2}$$

where e is the electron charge [28]. The electron concentration in pristine MoS_2 was 6.47×10^{11} cm^{-2}, whereas the hole concentration in P-doped MoS_2 was 1.01×10^{11} cm^{-2}. Based on these results, the complete p-type conduction of MoS_2 with the addition of P_2O_5 was demonstrated in this study.

4. Conclusions

We have demonstrated the p-type conduction of P-doped MoS_2 by P_2O_5 via a CVD process. Based on AFM and Raman measurements, pristine and P-doped MoS_2 were confirmed to have monolayer thickness with grain sizes in the order of 15 μm. From XPS data, it was suggested that the Fermi level of P-doped MoS_2 shifted by about 0.5 eV toward the valence band compared to the pristine MoS_2. FETs with P-doped monolayer MoS_2 showed p-type conduction with a field-effect mobility of 0.023 cm^2/V·s and an on/off current ratio of 10^3, while pristine MoS_2 FETs had n-type behavior with a field-effect mobility of 29.7 cm^2/V·s and an on/off current ratio of 10^5. The carriers in the FET channel were identified to be holes with a concentration of 1.01×10^{11} cm^{-2} in P-doped MoS_2 and electrons with a concentration of 6.47×10^{11} cm^{-2} in the pristine material. This phosphorous doping technique should be applicable to other TMD materials.

Author Contributions: J.S.L. performed the experiment, data analysis, discussed the results and wrote the paper; C.-S.P., T.Y.K. and Y.S.K. discussed the results and analyzed the data; and E.K.K. performed paper editing and supervision.

Funding: This research was supported by a National Research Foundation of Korea (NRF) grant funded by the Korean government (MSIP) (NRF-2016R1A2B4011706, NRF-2018R1A2A3074921).

Conflicts of Interest: The authors declare no conflict of interest.

References

1. Novoselov, K.S.; Geim, A.K.; Morozov, S.V.; Jiang, D.; Zhang, Y.; Dubonos, S.V.; Griorieva, I.V.; Firsov, A.A. Electric field effect in atomically thin carbon films. *Science* **2004**, *306*, 666–669. [CrossRef] [PubMed]
2. Bolotin, K.I.; Sikes, K.J.; Jiang, Z.; Klima, M.; Fudenberg, G.; Hon, J.; Kim, P.; Stormer, H.L. Ultrahigh electron mobility in suspended graphene. *Solid State Commun.* **2008**, *146*, 351–355. [CrossRef]
3. Liu, M.; Yin, X.; Ulin-Avila, E.; Geng, B.; Zentgraf, T.; Ju, L.; Wang, F.; Zhang, X. A graphene based broadband optical modulator. *Nature* **2011**, *474*, 64–67. [CrossRef] [PubMed]
4. Ozlem, S.; Akkaya, U. Graphene electronics: Thinking outside the silicon box. *Nat. Nanotechnol.* **2009**, *131*, 48–49.
5. Osada, M.; Sasaki, T. 2D inorganic nano-sheets: Two-dimensional dielectric nano-sheets: Novel nanoelectronics from nanocrystal building blocks. *Adv. Mater.* **2012**, *24*, 210–228. [CrossRef] [PubMed]
6. Gordon, R.; Yang, D.; Crozier, E.; Jiang, D.; Frindt, R. Structures of exfoliated single layers of WS_2, MoS_2, and $MoSe_2$ in aqueous suspension. *Phys. Rev. B* **2002**, *65*, 125407. [CrossRef]
7. Chu, D.; Pak, S.W.; Kim, E.K. Locally Gated SnS_2/hBN Thin Film Transistors with a Broadband Photoresponse. *Sci. Rep.* **2018**, *8*, 10585–10593. [CrossRef] [PubMed]
8. Lee, S.K.; Chu, D.; Yoo, J.; Kim, E.K. Formation of transition metal dichalcogenides thin films with liquid phase exfoliation technique and photovoltaic applications. *Sol. Energy Mater. Sol. Cells* **2018**, *184*, 9–14. [CrossRef]
9. Qiu, D.; Lee, D.U.; Pak, S.W.; Kim, E.K. Structural and optical properties of MoS_2 layers grown by successive two-step chemical vapor deposition method. *Thin Solid Films* **2015**, *587*, 47–51. [CrossRef]
10. Radisavljevic, B.; Radenovic, A.; Brivio, J.; Giacometti, V.; Kis, A. Single-layer MoS_2 transistors. *Nat. Nanotechnol.* **2011**, *6*, 147–150. [CrossRef]
11. Laskar, M.R.; Nath, D.N.; Ma, L.; Lee, E.W.; Lee, C.H.; Kent, T.; Yang, Z.; Mishra, R.; Roldan, M.A.; Idrobo, J.-C.; et al. p-type doping of MoS_2 thin films using Nb. *Appl. Phys. Lett.* **2014**, *104*, 092104. [CrossRef]
12. Momose, T.; Nakamura, A.; Daniel, M.; Shimomura, M. Phosphorous doped p-type MoS_2 polycrystalline thin films via direct sulfurization of Mo film. *AIP Adv.* **2018**, *8*, 025009. [CrossRef]
13. Dolui, K.; Rungger, I.; Pemmaraju, C.D.; Sanvito, S. Possible doping strategies for MoS_2 monolayers: Anab initiostudy. *Phys. Rev. B* **2013**, *88*, 075429. [CrossRef]

14. Xu, E.Z.; Liu, H.M.; Park, K.; Li, Z.; Losovyj, Y.; Starr, M.; Werbianskyj, M.; Fertig, H.A.; Zhang, S.X. p-Type transition-metal doping of large-area MoS_2 thin films grown by chemical vapor deposition. *Nanoscale* **2017**, *9*, 3576–3584. [CrossRef] [PubMed]

15. Zhang, K.; Bersch, B.M.; Joshi, J.; Addou, R.; Cormier, C.R.; Zhang, C.; Xu, K.; Briggs, N.C.; Wang, K.; Subramanian, S.; et al. Tuning the Electronic and Photonic Properties of Monolayer MoS_2 via In Situ Rhenium Substitutional Doping. *Adv. Funct. Mater.* **2018**, *28*, 1706950. [CrossRef]

16. Zhang, K.; Feng, S.; Wang, J.; Azcatl, A.; Lu, N.; Addou, R.; Wang, N.; Zhou, C.; Lerach, J.; Bojan, V.; et al. Manganese Doping of Monolayer MoS_2: The Substrate Is Critical. *Nano Lett.* **2015**, *15*, 6586–6591. [CrossRef]

17. Xie, Y.; Wang, Z.; Zhan, Y.; Zhang, P.; Wu, R.; Jiang, T.; Wu, S.; Wang, H.; Zhao, Y.; Nan, T.; et al. Controllable growth of monolayer MoS_2 by chemical vapor deposition via close MoO_2 precursor for electrical and optical applications. *Nanotechnology* **2017**, *28*, 084001. [CrossRef]

18. Wang, S.; Rong, Y.; Fan, Y.; Pacios, M.; Bhaskaran, H.; He, K.; Warner, J.H. Shape Evolution of Monolayer MoS_2 Crystals Grown by Chemical Vapor Deposition. *Chem. Mater.* **2014**, *26*, 6371–6379. [CrossRef]

19. Li, H.; Zhang, Q.; Yap, C.C.R.; Tay, B.K.; Edwin, T.H.T.; Olivier, A.; Baillargeat, D. From Bulk to Monolayer MoS_2: Evolution of Raman Scattering. *Adv. Funct. Mater.* **2012**, *22*, 1385–1390. [CrossRef]

20. Lee, C.G.; Yan, H.; Brus, L.E.; Heinz, T.F.; Hone, J.; Ryu, S. Anomalous Lattice Vibrations of Single-and Few-Layer MoS_2. *ACS Nano* **2010**, *4*, 2695–2700. [CrossRef]

21. Azcatl, A.; Qin, X.; Prakash, A.; Zhang, C.; Cheng, L.; Wang, Q.; Lu, N.; Kim, M.J.; Kim, J.; Cho, K.; et al. Covalent Nitrogen Doping and Compressive Strain in MoS_2 by Remote N_2 Plasma Exposure. *Nano Lett.* **2016**, *16*, 5437–5443. [CrossRef] [PubMed]

22. Chakraborty, A.; Bera, A.; Muthu, D.V.S.; Bhowmick, S.; Waghmare, U.V.; Sood, A.K. Symmetry-dependent phonon renormalization in monolayer MoS_2 transistor. *Phys. Rev. B* **2012**, *85*, 161403. [CrossRef]

23. Kukucska, G.; Koltai, J. Theoretical Investigation of Strain and Doping on the Raman Spectra of Monolayer MoS_2. *Phys. Status Solidi (B)* **2017**, *254*, 1700184. [CrossRef]

24. Suh, J.; Park, T.E.; Lin, D.Y.; Fu, D.; Park, J.; Jung, H.J.; Chen, Y.; Ko, C.; Jang, C.; Sun, Y.; et al. Doping against the native propensity of MoS_2: Degenerate hole doping by cation substitution. *Nano Lett.* **2014**, *14*, 6976–6982. [CrossRef] [PubMed]

25. Velicky, M.; Bissett, M.A.; Woods, C.R.; Toth, P.S.; Georgiou, T.; Kinloch, I.A.; Novoselov, K.S.; Dryfe, R.A. Photoelectrochemistry of Pristine Mono- and Few-Layer MoS_2. *Nano Lett.* **2016**, *16*, 2023–2032. [CrossRef] [PubMed]

26. McDonnell, S.; Addou, R.; Buie, C.; Wallace, R.M.; Hinkle, C.L. Defect-Dominated Doping and Contact Resistance in MoS_2. *ACS Nano* **2014**, *8*, 2880–2888. [CrossRef] [PubMed]

27. Zhang, Y.; Li, H.; Wang, H.; Xie, H.; Liu, R.; Zhang, S.L.; Qiu, Z.J. Thickness Considerations of Two-Dimensional Layered Semiconductors for Transistor Applications. *Sci. Rep.* **2016**, *6*, 29615–29622. [CrossRef] [PubMed]

28. Zhang, S.; Hill, H.M.; Moudgil, K.; Richter, C.A.; Walker, A.R.H.; Barlow, S.; Marder, S.R.; Hacker, C.A.; Pookpanratana, S.J. Controllable, Wide-Ranging n-Doping and p-Doping of Monolayer Group 6 Transition-Metal Disulfides and Diselenides. *Adv. Mater.* **2018**, *30*, 1802991–1802999. [CrossRef] [PubMed]

 *nanomaterials*

Article

1.34 µm Q-Switched Nd:YVO$_4$ Laser with a Reflective WS$_2$ Saturable Absorber

Taijin Wang [1], Yonggang Wang [1,*], Jiang Wang [1], Jing Bai [2], Guangying Li [3], Rui Lou [3] and Guanghua Cheng [4]

[1] School of Physics and Information Technology, Shaanxi Normal University, Xi'an 710119, China
[2] Department of Physics, Taiyuan Normal University, Taiyuan 030031, China
[3] State Key Laboratory of Transient Optics and Photonics, Xi'an Institute of Optics and Precision Mechanics, Chinese Academy of Sciences, Xi'an 710119, China
[4] Electronic Information College, Northwestern Polytechnical University, Xi'an 710072, China
* Correspondence: chinawygxjw@snnu.edu.cn

Received: 4 July 2019; Accepted: 22 August 2019; Published: 26 August 2019

Abstract: In this work, a Tungsten disulfide (WS$_2$) reflective saturable absorber (SA) fabricated using the Langmuir–Blodgett technique was used in a solid state Nd:YVO$_4$ laser operating at 1.34 µm. A Q-switched laser was constructed. The shortest pulse width was 409 ns with the repetition rate of 159 kHz, and the maximum output power was 338 mW. To the best of our knowledge, it is the first time that short laser pulses have been generated in a solid state laser at 1.34 µm using a reflective WS$_2$ SA fabricated by the Langmuir–Blodgett method.

Keywords: WS$_2$; saturable absorbers; Langmuir–Blodgett technique; Q-switched laser

---

1. Introduction

Saturable absorbers (SA) have been used as a switching element to generate short pulses in passively Q-switched lasers. It is mainly represented by transition metal ions–doped bulk crystals like Cr^{4+}:YAG and V^{3+}:YAG [1–5], semiconductor materials like the Semiconductor Saturable Absorbing Mirror (SESAM) [6–9], and two-dimensional (2D) materials [10–17].

The fabrication method of switching elements is very important and determines the performances of the Q-switching lasers. The Langmuir–Blodgett (LB) technique is a convenient and low-cost method for preparing ultrathin nano materials films [18].

Two-dimensional materials have been widely used in laser applications [19–23] due to their simple structure and remarkable wide spectral band [24–26]. 2D atomically thin Tungsten disulfide (WS$_2$) nanosheets exfoliated from bulk counterparts have shown exotic electronic and optical properties, such as indirect-to-direct bandgap transition with a reducing number of layers (the indirect band gap is ~1.3 eV and the direct band gap of its monolayer form is up to 2.1 eV), high carrier mobility, and strong spin–orbit coupling due to their broken inversion symmetry, which have enabled wide potential applications in viable photonic and optoelectronic devices [27–29]. As a kind of 2D material, WS$_2$ has been successfully developed to produce short pulses in lasers with various wavelengths such as 1.06 µm, 1.53 µm, 1.65 µm, and 3 µm [30–34].

In this paper, the LB technique was used to coat few-layer WS$_2$ onto a silver-coated mirror. In this way, a low-cost reflective WS$_2$ saturable absorber (SA) was prepared. Based on the reflective WS$_2$ SA, a passive Q-switched solid state Nd:YVO$_4$ laser was constructed, which generated short pulses at 1.34 µm. The maximum average Q-switched output power of 338 mW was obtained with the pulse repetition rate of 159 kHz, corresponding to the single pulse energy of 2.13 µJ and peak power of 5.20 W, respectively. The results indicate that the WS$_2$ can be fabricated by the Langmuir–Blodgett method and used as a Q-switch element in solid state lasers to generate short pulses at 1.34 µm.

2. Materials and Methods

2.1. WS$_2$ Saturable Absorber Fabrication

The few-layer WS$_2$ suspension was fabricated from the bulk WS$_2$ by liquid phase exfoliation. A bulk WS$_2$ (from XF NANO Inc., Nanjing, China) was ultrasonicated for 24 h and centrifuged for 20 min to get the aqueous solution with the concentration of 2 mg/mL.

The methanol, chloroform, and as-prepared WS$_2$ supernatant with the volume ratio of 1:1:4 was prepared and ultrasonicated for 15 min. The Raman spectrum of the WS$_2$ silicon wafer was measured by a Raman spectrometer (LabRam confocal Microprobe system, Horiba Jobin Yvon, Paris, France).

The reflective WS$_2$ saturable absorber, a silver mirror coated with WS$_2$ saturable absorber by the Langmuir–Blodgett technique, and the Langmuir–Blodgett system (JML04C1, 2017JM7085, Powereach, Shanghai, China), are shown in Figure 1. The silver mirror was composed of a 180 nm silver film and a 20 nm silica protection film evaporated on a quartz plate by the electron beam aided evaporation technique.

The prepared WS$_2$ solution was dipped into a trough containing deionized water and spread on the surface of the cell. The trough contained 200 mL deionized water and the pH of deionized water was 7.0. The instillation stopped until the pressure of the surface, measured by a force transducer, reached 35 mN/m steady. Then the silver mirror, pre-inserted into the deionized water, was pulled up slowly. At the same time, the surface of the liquid was compressed by two mobile barriers with the speed of 4.85 mm/min under the control of a motor. After the silver mirror coating, the WS$_2$ films were pulled out from the liquid completely and then dried at 80 °C for 10 min. The reflective WS$_2$ SA was fabricated successfully.

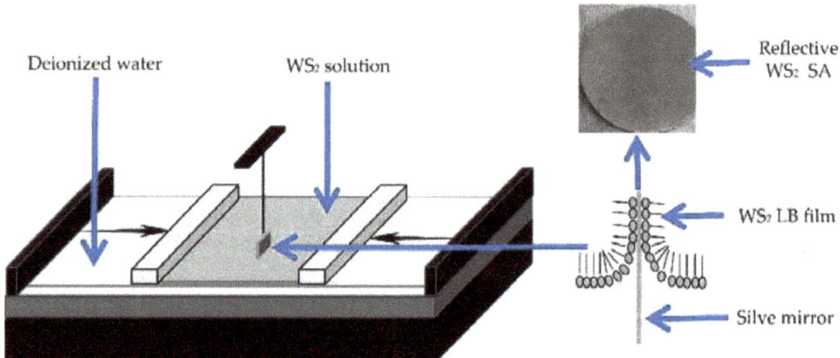

Figure 1. The fabrication of the reflective WS$_2$ saturable absorber (SA) by the Langmuir–Blodgett (LB) system. Inset: the reflective WS$_2$ saturable absorber.

2.2. Characterization of WS$_2$ Saturable Absorber

The surface of the WS$_2$ films was characterized by a scanning electron microscope (SEM, Nova NanoSEM Training-X50 series, FEI, Eindhoven, The Netherlands) and the thickness of the WS$_2$ films was characterized by an atomic force microscope (AFM, Dimension Icon, Bruker Nano Inc., Mannheim, Germany).

A spectrophotometer (Perkin-Elmer, UV-Lambda 1050, Downers Grove, IL, USA) was used to measure the linear optical reflectivity curve of the reflective WS$_2$ saturable absorber and the nonlinear optical characteristics of the reflective WS$_2$ SA were measured by a balanced twin-detector measurement technique, which was described in [35]. The pump source for the nonlinear optical measurement was a self-made acoustic-optically Q-switched Nd:YVO$_4$ laser with the pulse of 40 ns and a repetition rate of 10 kHz at 1.34 μm.

2.3. Laser Cavity

Figure 2 shows the schematic setup of the Nd:YVO$_4$ passively Q-switched laser with a reflective WS$_2$ SA at 1.34 µm.

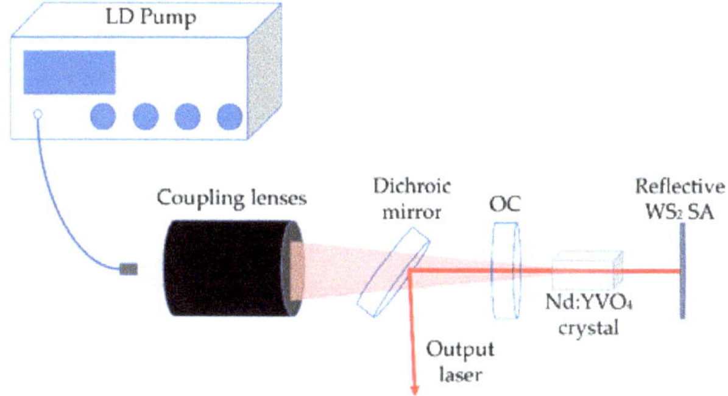

Figure 2. The schematic setup of the Nd:YVO$_4$ passively Q-switched laser.

There was a $3 \times 3 \times 10$ mm a-cut Nd:YVO$_4$ crystal with a Nd^{3+} ions doping concentration of 0.5 at.%, and the 808 nm anti-reflective films were coated onto both sides of its ends. Water-cooled equipment was used to maintain the temperature of the laser crystal at 12 °C. The crystal was wrapped with indium foils contacted tightly with copper heat sink.

A fiber-coupled laser diode (LD) with the maximum output power of 50 W and the central wavelength of 808 nm was used as the pump source. The pump light was focused on the Nd:YVO$_4$ crystal with a pump spot diameter of 400 µm after passing through a 1:1 coupling lens, a flat mirror, and a concave output coupler (OC). The flat mirror was coated with anti-reflective film at 808 nm and high-reflective film (R > 99.9%) at 1342 nm. The output coupler with the curvature radius of r = 100 mm had a transmission of 5%. The length of the cavity was about 14 mm. it was set up with a reflective WS$_2$ saturable absorber and an output coupler. The distance from the laser crystal to the output coupler and the reflective WS$_2$ saturable absorber were 1 and 3 mm, respectively.

The average output power of the Q-switched laser at 1.34 µm can be measured accurately by a power meter. The data of the output Q-switched pulse repetition rate and duration were recorded by a digital oscilloscope (Rohde & Schwarz, RTO1014, Munich, Germany) with a photodetector (Thorlabs, DET08C/M, Munich, Germany). A laser spectrum analyzer (YOKOGAWA, AQ6370D, Suzhou, China) was employed to record the spectrum.

3. Results and Discussion

3.1. Characteristics of WS$_2$ Saturable Absorber

Figure 3 shows the Raman spectrum of the few-layer WS$_2$ excited by a 532 nm laser source. The locations of two characteristic Raman active vibration modes, viz., E^1_{2g} (in-plane) at 356.3 cm^{-1} and A^1_g (out-of-plane) at 417.0 cm^{-1}, should be in agreement with other reported few-layer WS$_2$ [36].

The thickness and surface roughness of the WS$_2$ SA films are shown in Figure 4. The image of the atomic force microscope (AFM) is shown in Figure 4a. The thickness of the WS$_2$ films is about 5 nm, and the surface roughness is less than 1 nm in Figure 4b. In addition, the surface of the WS$_2$ films is shown by the scanning electron microscope (SEM) image in Figure 4c. In short, it was determined that the surface of WS$_2$ films was very uniformed. Moreover, we could estimate the size of the particle of the WS$_2$ films, it was about dozens of micron.

The reflection spectrum of the reflective WS$_2$ saturable absorber is measured by a wavelength range from 1000 to 1400 nm in Figure 5. It shows the reflectivity of the sample is about 64.8%.

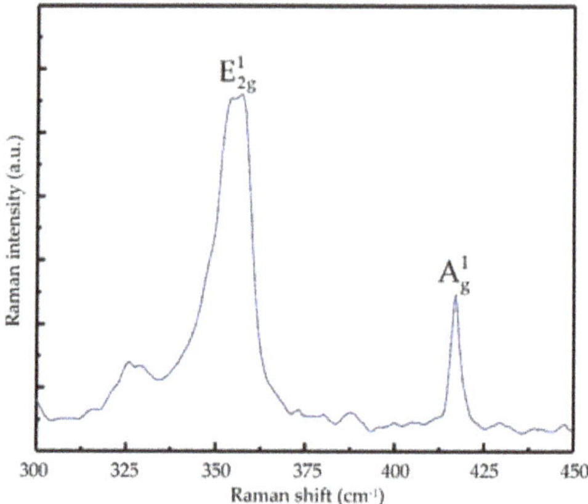

Figure 3. The Raman spectrum.

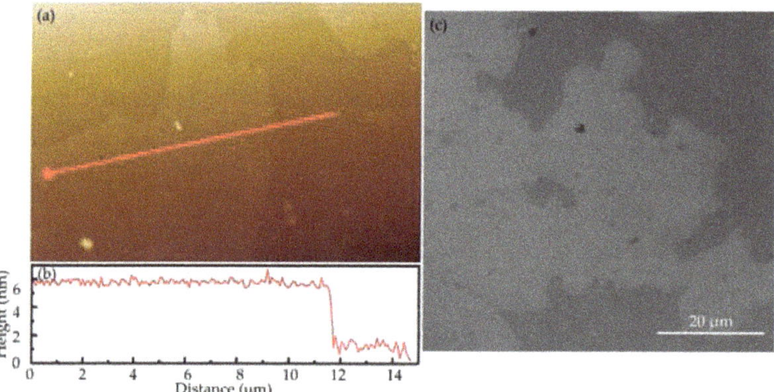

Figure 4. (**a**) Image of the atomic force microscope (AFM); (**b**) the thickness of the WS_2 films; (**c**) image of the scanning electron microscope (SEM).

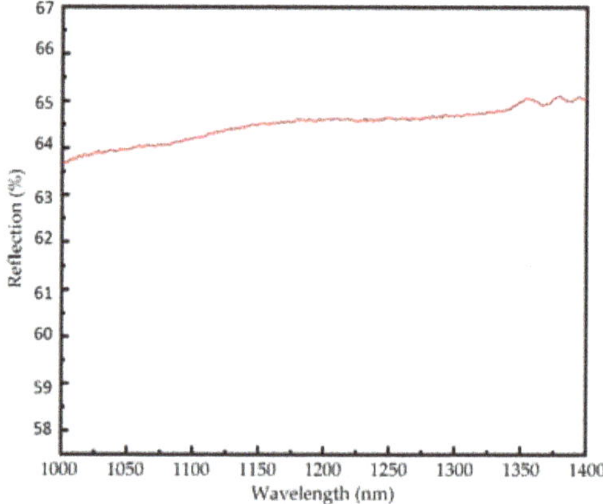

Figure 5. The reflectance spectrum of the reflective WS_2 saturable absorber.

The nonlinear absorption saturation characteristics of the WS_2 saturable absorber is displayed in Figure 6. The schematic diagram of nonlinear optical absorption measurement is depicted as the inset.

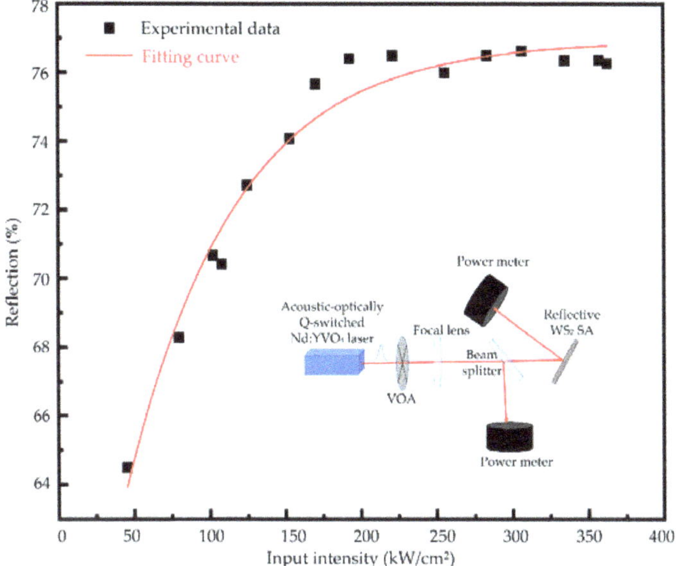

Figure 6. The nonlinear absorption saturation characteristics. Inset: the schematic diagram of nonlinear optical absorption measurement.

The reflectivity of the WS_2 saturable absorber versus different incident pulse energy intensities was recorded. The data of the reflectivity is depicted as dots in Figure 7, and fitted by the following equation [35]: $T(I) = 1 - \Delta T \exp(-I/Isat) - Tns$, where $T(I)$ is the reflectivity of the reflective WS_2 saturable absorber, ΔT is the modulation depth, I_{sat} is the saturable intensity, and T_{ns} is the non-saturable loss. The modulation depth and the saturation intensity of the reflective WS_2 SA were simulated to be 24.5% and 71.9 kW/cm^2, respectively.

3.2. WS_2 Q-Switched Laser

Firstly, the operation of the continuous wave (CW) Nd:YVO$_4$ laser with a output coupler and a high-reflective mirror was investigated. The relationship between continuous wave laser output power and pump power is observed in Figure 7a. As shown in Figure 7a, the pump power threshold of the continuous wave laser and the slope efficiency of the almost linear relationship are 37 mW and 22.8%, respectively. No self-Q-switched pulse was observed in the generation of the continuous wave laser.

The operation of the passively Q-switched (QW) laser was effected after replacing the HR mirror with the reflective WS_2 saturable absorber. The data of the Q-switched average output power are shown in Figure 7a. The Q-switched operation remained unchanged when the pump power increased from 1.84 to 2.83 W, and the Q-switched laser output power changed from 132 to 338 mW correspondingly, with a slope efficiency of 19.9%.

The pulse widths and repetition rates were recorded synchronously. The evolution of the pulse width and repetition rates of the pump power is presented in Figure 7b. Based on the data of the output Q-switched laser, it was directed to get the single pulse energies and peak powers. As displayed in Figure 7c, the maximum single pulse energy of 2.13 µJ and pulse peak power of 5.20 W was obtained when the pump power was 2.83 W.

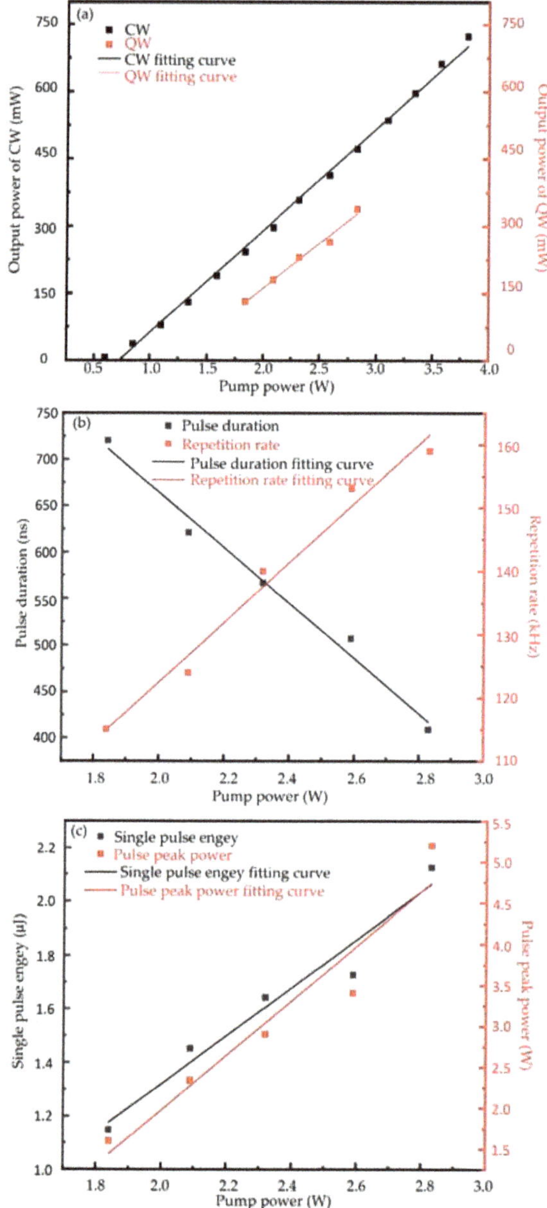

Figure 7. (**a**) The average output power of the continuous wave (CW) laser and of the Q-switched (QW) laser versus the pump power; (**b**) evolutions of the pulse duration and the pulse repetition rate with the pump power; (**c**) evolutions of the single pulse energy and the pulse peak power with the pump power.

Three different individual pulses under different pump powers were depicted in Figure 8a,b to show the evolution of the pulse width and repetition rates with the pump power visually. It demonstrated that the pulse width decreased from 720 to 409 ns of and the repetition rate increased from 115 to 159 kHz with the increase of the pump power from 1.84 to 2.83 W. A slight jitter of the pulse trains is observed from Figure 8a. The jitter was primarily caused by the thermal instability of the reflective SA under long time laser illumination. It was also possible that the instability of the laser cavity attributed to the thermal lens effect from the laser crystal to give rise to the jitter. The shortest pulse duration of 409 ns was obtained and is displayed in Figure 8b. The QW spectrum was measured and is shown in Figure 8c. The central wavelength (λ_c) was 1342 nm with the bandwidth of 0.12 nm.

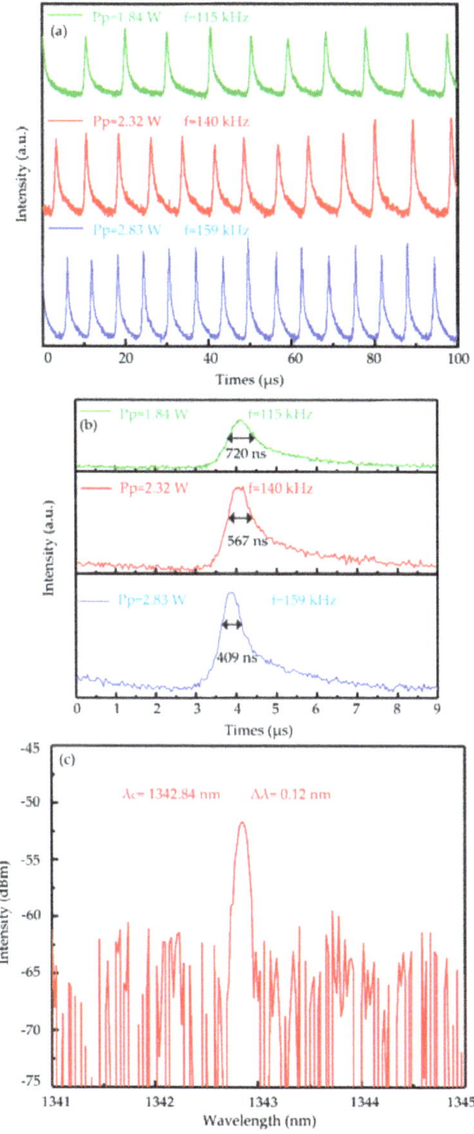

Figure 8. (**a**) The pulse trains of Q-switched lasers under different pump power; (**b**) the individual pulse under different pump power; (**c**) the Q-switched laser spectrum.

4. Conclusions

In this work, we presented a new kind of reflective WS_2 saturable absorber fabricated using the Langmuir–Blodgett technique and constructed, for the first time, a passively Q-switched $Nd:YVO_4$ solid state laser at 1.3 µm with the absorber. It had ideal characteristics in thickness and the uniformity of the nanomaterials. The shortest duration was achieved with pulse width of 409 ns, and the highest peak power was 5.20 W. These results indicate that a reflective WS_2 saturable absorber with perfect characteristics fabricated using the Langmuir–Blodgett technique can be a promising optical modulator to generate short pulses at 1.3 µm.

Author Contributions: Conceptualization, T.W.; Data curation, T.W., J.W., J.B., G.L. and R.L.; Funding acquisition, Y.W.; Investigation, T.W. and J.W.; Methodology, T.W., J.W. and G.C.; Project administration, Y.W.; Supervision, T.W.; Writing—original draft, T.W.; Writing—review & editing, T.W. and Y.W.

Funding: This research was funded by Central University Special Fund Basic Research and Operating, grant number GK201702005.

Conflicts of Interest: The authors declare no conflict of interest.

References

1. Cheng, L.; Shengzhi, Z.; Guiqiu, L.; Kejian, Y.; Dechun, L.; Tao, L.; Wenchao, Q.; Tianli, F.; Xintian, C.; Xiaodong, X.; et al. Experimental and theoretical study of a passively Q-switched Nd:LuAG laser at 1.3 µm with a V^{3+}:YAG saturable absorber. *Opt. Soc. Am.* **2015**, *32*, 1001–1006.

2. Zayhowski, J.J.; Dill, I.C. Diode-pumped passively Q-switched picosecond microchip lasers. *Opt. Lett.* **1994**, *19*, 1427–1429. [CrossRef] [PubMed]

3. Zhang, X.; Zhao, S.; Wang, Q.; Zhang, Q.; Sun, L. Optimization of Cr^{4+}-doped saturable-absorber Q-switched lasers. *IEEE J. Quantum Electron.* **1997**, *33*, 2286–2294. [CrossRef]

4. Sandu, O. High-peak power, passively Q-switched, composite, all-polycrystalline ceramic $Nd:YAG/Cr^{4+}$: YAG lasers. *Quantum Electron.* **2012**, *42*, 211–215. [CrossRef]

5. Chen, Y.F.; Chang, C.C.; Lee, C.Y.; Sung, C.L.; Tung, J.C.; Su, K.W.; Liang, H.C.; Chen, W.D.; Zhang, G. High-peak-power large-angular-momentum beams generated from passively Q-switched geometric modes with astigmatic transformation. *Photonics Res.* **2017**, *5*, 561–566. [CrossRef]

6. Kajava, T.T.; Gaeta, A.L. Q switching of a diode-pumped Nd:YAG laser with GaAs. *Opt. Lett.* **1996**, *21*, 1244–1246. [CrossRef] [PubMed]

7. Spühler, G.J.; Paschotta, R.; Fluck, R.; Braun, B.; Moser, M.; Zhang, G.; Gini, E.; Keller, U. Experimentally confirmed design guidelines for passively Q-switched microchip lasers using semiconductor saturable absorbers. *J. Opt. Soc. Am. B* **2001**, *18*, 886. [CrossRef]

8. Chen, J.; Lin, J.T.; Hung, T.C. Q-switched mode-lock pulses laser range finder. *Proc. SPIE Int. Soc. Optic. Eng.* **2005**, *5627*, 270–277.

9. Nikkinen, J.; Harkonen, A.; Leino, I.; Guina, M. Generation of sub-100 ps pulses at 532 nm, 355 nm and 266 nm using a SESAM Q-switched microchip laser. *IEEE Photonics Technol. Lett.* **2017**, *29*, 1816–1819. [CrossRef]

10. Popa, D.; Sun, Z.; Hasan, T.; Torrisi, F.; Wang, F.; Ferrari, A.C. Graphene q-switched, tunable fiber laser. *Appl. Phys. Lett.* **2011**, *98*, 435. [CrossRef]

11. Xu, J.; Li, X.; He, J.; Hao, X.; Yang, Y.; Wu, Y.; Liu, S.; Zhang, B. Efficient graphene Q switching and mode locking of 1.34 µm neodymium lasers. *Opt. Lett.* **2012**, *37*, 2652–2654. [CrossRef]

12. Cheng, C.; Liu, H.; Tan, Y.; Tan, Y.; de Aldana, J.R.V.; Chen, F. Passively q-switched waveguide lasers based on two-dimensional transition metal diselenide. *Opt. Express* **2016**, *24*, 10385–10390. [CrossRef]

13. Lin, M.; Peng, Q.; Hou, W.; Fan, X.; Liu, J. 1.3 lm Q-switched solid-state laser based on few-layer ReS2 saturable absorber. *Optic. Laser Technol.* **2019**, *109*, 90–93. [CrossRef]

14. Ismail, E.I.; Kadir, A.; Latiff, A.A.; Ahmad, H.; Harun, S.W. Q-switched erbium-doped fiber laser operating at 1502nm with molybdenum disulfide saturable absorber. *J. Nonlinear Optic. Phys. Mater.* **2016**, *25*, 1650025. [CrossRef]

15. Wang, K.; Yang, K.; Zhang, X.; Zhao, S.; Luan, C.; Liu, C.; Wang, J.; Xu, X.; Xu, J. Passively Q-switched laser at 1.3 µm with few-layered MoS2 saturable absorber. *IEEE J. Sel. Top. Quantum Electron.* **2016**, *23*, 1077–1260.

16. Chu, Z.; Liu, J.; Guo, Z.; Zhang, H. 2 µm passively q-switched laser based on black phosphorus. *Opt. Mater. Express* **2016**, *6*, 2374. [CrossRef]

17. Fauziah, C.M.; Rosol, A.H.A.; Latiff, A.A.; Harun, S.W. The generation of q-switched erbium-doped fiber laser using black phosphorus saturable absorber with 8% modulation depth. *IOP Conf. Ser. Mater. Sci. Eng.* **2017**, *210*, 012043. [CrossRef]

18. Jiang, W.; Yonggang, W.; Taijin, W.; Guangying, L.; Rui, L.; Guanghua, C.; Jing, B. Nonlinear Optical Response of Graphene Oxide Langmuir-Blodgett Film as Saturable Absorbers. *Nanomaterials* **2019**, *9*, 640.

19. Jhon, Y.; Koo, J.; Anasori, B.; Seo, M.; Lee, J.H.; Gogotsi, Y.; Jhon, Y.M. Metallic MXene Saturable Absorber for Femtosecond Mode-Locked Lasers. *Adv. Mater.* **2017**, *29*, 1702496. [CrossRef]

20. Li, L.; Lv, R.; Wang, J.; Chen, Z.; Wang, H.; Liu, S.; Ren, W.; Liu, W.; Wang, Y. Optical Nonlinearity of ZrS2 and Applications in Fiber Laser. *Nanomaterials* **2019**, *9*, 315. [CrossRef]

21. Tuo, M.; Xu, C.; Mu, H.; Bao, X.; Wang, Y.; Xiao, S.; Ma, W.; Li, L.; Tang, D.; Zhang, H.; et al. Ultrathin 2D transition metal carbides for ultrafast pulsed fiber lasers. *ACS Photonics* **2018**, *5*, 1808–1816. [CrossRef]

22. Wang, F. Two-dimensional materials for ultrafast lasers. *Chin. Phys. B* **2017**, *26*, 034202. [CrossRef]

23. Li, D.; Xue, H.; Qi, M.; Wang, Y.; Aksimsek, S.; Chekurov, N.; Kim, C.; Li, C.; Riikonen, J.; Ye, F.; et al. Graphene actively q-switched lasers. *2D Materials* **2017**, *4*, 025095. [CrossRef]

24. He, J.; Tao, L.; Zhang, H.; Zhou, B.; Li, J. Emerging 2D materials beyond graphene for ultrashort pulse generation in fiber lasers. *Nanoscale* **2019**, *11*, 2577–2593. [CrossRef]

25. Yu, S.; Wu, X.; Wang, Y.; Guo, X.; Tong, L. 2d materials for optical modulation: Challenges and opportunities. *Adv. Mater.* **2017**, *29*, 1606128. [CrossRef]

26. Liu, X.; Guo, Q.; Qiu, J. Emerging Low-Dimensional Materials for Nonlinear Optics and Ultrafast Photonics. *Adv. Mater.* **2017**, *29*, 1605886. [CrossRef]

27. Gutiérrez, H.R.; Perea-López, N.; Elías, A.L.; Berkdemir, A.; Wang, B.; Lv, R.; López-Urías, F.; Crespi, V.H.; Terrones, H.; Terrones, M. Extraordinary Room-Temperature Photoluminescence in Triangular WS2 Monolayers. *Nano Lett.* **2013**, *13*, 3447–3454. [CrossRef]

28. Zhao, W.; Ghorannevis, Z.; Chu, L.; Toh, M.; Kloc, C.; Tan, P.-H.; Eda, G. Evolution of Electronic Structure in Atomically Thin Sheets of WS2 and WSe2. *ACS Nano* **2013**, *7*, 791–797. [CrossRef]

29. Wu, K.; Zhang, X.; Wang, J.; Li, X.; Chen, J. WS2 as a saturable absorber for ultrafast photonic applications of mode-locked and Q-switched lasers. *Opt. Express* **2015**, *23*, 11453. [CrossRef]

30. Tang, W.; Wang, Y.; Yang, K.; Zhao, J.; Zhao, S.; Li, G.; Li, D.; Li, T.; Qiao, W. 1.36 W Passively Q-Switched YVO4/Nd:YVO4 Laser With a WS2 Saturable Absorber. *IEEE Photonics Technol. Lett.* **2017**, *29*, 470–473. [CrossRef]

31. Tang, C.Y.; Cheng, P.K.; Tao, L.; Long, H.; Tsang, Y.H. Passively Q-switched Nd:YVO$_4$ laser using WS$_2$ saturable absorber fabricated by radio frequency magnetron sputtering deposition. *J. Lightwave Technol.* **2017**, *35*, 4120–4124. [CrossRef]

32. Wenjun, L.; Lihui, P.; Hainian, H.; Zhongwei, S.; Ming, L.; Hao, T.; Zhiyi, W. Dark solitons in WS$_2$ erbium-doped fiber lasers. *Photonics Res.* **2016**, *4*, 111–114.

33. Zhang, S.; Guo, L.; Fan, M.; Lou, F.; Zhao, S. Passively Q-Switched Er:LuAG Laser at 1.65 µm Using MoS$_2$ and WS$_2$ Saturable Absorbers. *IEEE Photonics J.* **2017**, *9*, 1–7.

34. Chen, W.; Luo, H.; Zhang, H.; Li, C.; Xie, J.; Li, J.; Liu, Y. Passively Q-switched mid-infrared fluoride fiber laser around 3 µm using a tungsten disulfide (WS$_2$) saturable absorber *Laser Phys. Lett.* **2016**, *13*, 105108.

35. Taijin, W.; Jiang, W.; Yonggang, W.; Xiguang, Y.; Sicong, L.; Ruidong, L.; Zhendong, C. High-power passively Q-switched Nd:GdVO$_4$ laser with a reflective graphene oxide saturable absorber. *Chin. Opt. Lett.* **2019**, *2*, 020009. [CrossRef]

36. Gang, Z.; Yonggang, W.; Zhiyong, J.; Dailin, L.; Zhendong, C. Tungsten disulfide saturable absorber for passively Q-Switched YVO4/Nd:YVO4/YVO4 laser at 1342.2 nm. *Opt. Mater.* **2019**, *92*, 95–99.

Article

Improvement of the Bias Stress Stability in 2D MoS$_2$ and WS$_2$ Transistors with a TiO$_2$ Interfacial Layer

Woojin Park [1,†], Yusin Pak [2,†], Hye Yeon Jang [1], Jae Hyeon Nam [1], Tae Hyeon Kim [1], Seyoung Oh [1], Sung Mook Choi [3], Yonghun Kim [3] and Byungjin Cho [1,*]

[1] Department of Advanced Material Engineering, Chungbuk National University, Chungdae-ro 1, Seowon-Gu, Cheongju, Chungbuk 28644, Korea
[2] Department of Nanobio Materials and Electronics, GIST, 123 Cheomdan-gwagiro, Buk-gu, Gwangju 61005, Korea
[3] Materials Center for Energy Department, Surface Technology Division, Korea Institute of Materials Science (KIMS), 797 Changwondaero, Sungsan-gu, Changwon, Gyeongnam 51508, Korea
* Correspondence: bjcho@chungbuk.ac.kr; Tel.: +82-(0)43-261-2417
† These authors contributed equally to this work.

Received: 11 July 2019; Accepted: 8 August 2019; Published: 12 August 2019

Abstract: The fermi-level pinning phenomenon, which occurs at the metal–semiconductor interface, not only obstructs the achievement of high-performance field effect transistors (FETs) but also results in poor long-term stability. This paper reports on the improvement in gate-bias stress stability in two-dimensional (2D) transition metal dichalcogenide (TMD) FETs with a titanium dioxide (TiO$_2$) interfacial layer inserted between the 2D TMDs (MoS$_2$ or WS$_2$) and metal electrodes. Compared to the control MoS$_2$, the device without the TiO$_2$ layer, the TiO$_2$ interfacial layer deposited on 2D TMDs could lead to more effective carrier modulation by simply changing the contact metal, thereby improving the performance of the Schottky-barrier-modulated FET device. The TiO$_2$ layer could also suppress the Fermi-level pinning phenomenon usually fixed to the metal–semiconductor interface, resulting in an improvement in transistor performance. Especially, the introduction of the TiO$_2$ layer contributed to achieving stable device performance. Threshold voltage variation of MoS$_2$ and WS$_2$ FETs with the TiO$_2$ interfacial layer was ~2 V and ~3.6 V, respectively. The theoretical result of the density function theory validated that mid-gap energy states created within the bandgap of 2D MoS$_2$ can cause a doping effect. The simple approach of introducing a thin interfacial oxide layer offers a promising way toward the implementation of high-performance 2D TMD-based logic circuits.

Keywords: MoS$_2$; WS$_2$; interfacial layer; contact resistance; bias stress stability

---

1. Introduction

The process of extreme scaling-down to reach a physical channel length limit of sub-100 nm has caused critical problems, such as a short channel effect and increased leakage current. To address these limitations, efforts have recently been made to scrutinize promising semiconducting materials. In particular, atomically thin layered transition metal dichalcogenides (TMDs) have attracted great attention due to their extraordinary electrical, optical, and mechanical properties [1–9]. One of their most attractive properties is the existence of a band-gap and its facile engineering. For instance, single-layer molybdenum disulfide (MoS$_2$) has a direct band-gap of ~1.8 eV, and multilayer MoS$_2$ has an indirect band-gap of ~1.2 eV [1]. The physical properties of 2D TMDs have led to their applications in various electronic devices such as transistors, memory devices, and opto-electronic devices [10–18]. Among them, the most promising device is the field caused effect transistor (FET), which functions as an essential switching component of display back-plane circuits [12].

However, a few challenging issues around employing 2D TMD-based FETs for practical applications have to be resolved. Fabricating large-scale, high-quality continuous 2D TMD films and the direct deposition of the gate dielectric layer on a 2D surface with a low surface energy are important issues in terms of the utilization of conventional Si fabrication infrastructures and the realization of high-mobility FETs. Furthermore, the unreliable performance of 2D TMD FETs has been a critical concern that must be preferentially addressed. Chemically and mechanically disordered surface and interface states are the origin of the performance instability of semiconductor devices, causing a large hysteresis window and a significant threshold voltage (V_{TH}) shift.

The passivation of the polymer layer on the 2D TMDs is an efficient countermeasure against the instability of 2D semiconductor-based FET performance [19,20]. Using a similar method, Zheng et al. reported that the hysteresis window of the 2D layered materials capped with an Al_2O_3 was considerably reduced [20]. Meanwhile, the contact engineering strategy for modifying the interface states between a metal and a 2D semiconductor has been actively studied [21–28]. Because the operation of the 2D TMD FET is based on a modulation of the Schottky-barrier, the interface quality at the metal/TMD contact becomes more critical. Several approaches to reduce the contact resistance, including a doping technique and selection of proper work function metal, have been proposed [26,28]. Meanwhile, Fermi-level pinning usually occurs at a metal/semiconductor contact region, causing high contact resistance due to a fixed high band offset regardless of the work function value of the metal [25,26,29]. Because the interface states usually serve as carrier trapping sites, it is hard to realize the high performance of 2D TMD FETs. Thus, a reliable and simple approach for Fermi-level depinning is necessary. The corresponding result was reported for an MoS_2 device with an interfacial oxide layer [29].

Herein, the effect of the interfacial buffer layer at the metal/2D TMD (MoS_2 and WS_2) contact on transistor performance was experimentally and theoretically investigated. Titanium dioxide (TiO_2) was used as a buffer layer because its band offset with MoS_2 and WS_2 is relatively small and tunnel resistance can be minimized with the thin TiO_2 layer. By employing a TiO_2 interlayer, interface states were successfully reduced, achieving an increased drive current and the enhancement of long term bias stress stability. In addition, the role of the TiO_2 layer on MoS_2 was theoretically elucidated using a density function theory (DFT) simulation. It can be highlighted that we suggested a facile approach to achieve both higher transistor performance and stability at the same time.

2. Materials and Methods

A mechanical exfoliation method using scotch tape to obtain high-quality 2D TMD flakes was adopted, and then the exfoliated 2D TMD flakes (MoS_2 and WS_2) were transferred onto a SiO_2 (300 nm)/heavily doped Si substrate. To identify the existence of the 2D TMDs, MoS_2 was mechanically exfoliated from the bulk mineral, and the multilayer MoS_2 was characterized using Raman spectroscopy (Figure 1a). LabRAM ARAMIS (laser wavelength: 473 nm, 50 mW) was used for Raman measurements. Two prominent peaks feature the in-plane E^1_{2g} mode (~384 cm^{-1}) and the out-of-plane A_{1g} mode (~409 cm^{-1}) of the MoS_2. A frequency difference of ~25 cm^{-1} between two vibrational modes indicates a multilayer MoS_2. To determine the thickness of the exfoliated MoS_2, we performed an atomic force microscopy (AFM) analysis. As shown in Figure 1b, the 92 nm-thick MoS_2 was transferred onto the SiO_2/Si substrate using a typical scotch-tape exfoliation method.

To investigate the effect of the TiO_2 interlayer on the device's contact properties, 2D FET devices with back gate electrodes were fabricated: a control device without TiO_2 and a testing device with TiO_2. Figure 1c shows the 3D schematic image of the FET device with the 2D TMD-TiO_2-Ti/Au structure. The TiO_2 interfacial layer on the 2D TMDs was deposited using an atomic layer deposition (ALD) technique based on a tetrakis-dimethyl-amido-titanium (TDMAT) precursor at 200 °C. The pulse and purging times were 0.2 s and 20 s, respectively. The number of cycles were 15, resulting in a 2~3 nm thickness. The thickness of the TiO_2 layer was also optimized to avoid high tunnel resistance. The 2D TMD transistor devices were made by a conventional photolithography process. Photolithography was conducted after spin-coating of the photoresist (AZ 5214, MicroChemicals, Germany), and the metal was deposited by a physical vapor evaporator. Electron beam evaporation was selected to minimize the physical damage on the surface of the TMDs. Lift-off processes were sequentially performed to make the source and drain electrodes. The channel distance between source and drain was ~3 µm. After device fabrication, the post-annealing process was conducted in a vacuum environment at 300 °C. The process of the vacuum annealing step included a 30 min ramping time to 300 °C, for a 1 h duration, and a cool down at room temperature. The electrical characterization (transfer, output, and stress measurement) was performed with a Keithley 4200-SCS (Keithley, Cleveland, OH, US). Stress measurement followed the conventional stress-measure-stress sequence for 10,000 s, which is summarized in Figure S6 of the Supplementary Materials information.

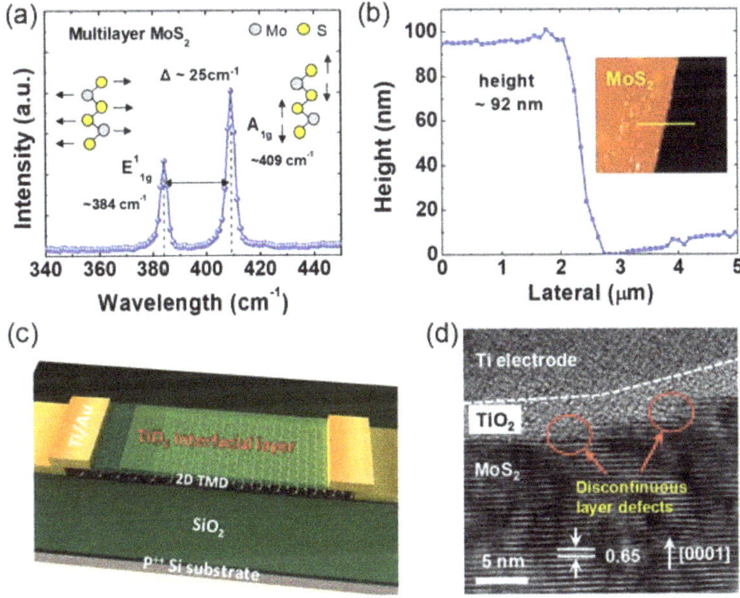

Figure 1. (**a**) Raman spectrum and (**b**) atomic force microscopy (AFM) analysis of a multilayer MoS$_2$; (**c**) 3D schematic image of a transition metal dichalcogenide field effect transistor (TMD FET) device; (**d**) a high-resolution transmission electron microscopy (HRTEM) image of the MoS$_2$-TiO$_2$-Ti stacked structure.

Figure 1d shows a cross-sectional high-resolution transmission electron microscopy (HRTEM) image of the MoS$_2$-TiO$_2$-Ti stacked structure. The lattice constant of the MoS$_2$ was measured to be ~0.65 nm along the c-plane [0001] direction in a hexagonal close-packed crystal structure. A thin (~3 nm-thick) TiO2 layer, deposited using the atomic layer deposition process, was inserted between the Ti metal and MoS$_2$. Interestingly, the discontinuous layers of the MoS$_2$ layers exhibited a step-like crystal structure. Thus, it is reasonably expected that randomness in the defect density for the exposed edge planes and basal planes can cause considerable deviation from the physical interface states, thereby inducing a large difference in the electrical properties of MoS$_2$. The structural disorder of the MoS$_2$ surface is also a strong source for Fermi-level pinning, which caused some points of the band gap to be locked (pinned) to the Fermi-level. This made the Schottky-barrier height considerably insensitive to the metal's work function. The Fermi-level pinning phenomenon, with respect to various metals (for instance, Ti, Cr, Au, and Pd), is illustrated in Figure S1 in the Supplementary Materials information. Even in the corresponding literature studies, the existence of dangling bonds in TMD has been proven via in-depth analyses, such as scanning tunneling microscopy and inductively coupled plasma-mass spectroscopy [30–33].

3. Results and Discussion

To investigate the influence of a TiO_2 interfacial layer on the MoS_2 and WS_2 device performance, electrical measurements were performed. Basic electrical characterizations were carried out with a Keithley 4200-SCS (Keithley, Cleveland, OH, US) analyzer. Figure 2a shows a comparison between the transfer characteristics (I_{DS}-V_{BG}) of the MoS_2-Ti and MoS_2-TiO_2-Ti devices. The gate-bias sweeping ranged from −50 to 20 V at a fixed drain voltage of 0.1 V. A typical unipolar n-type behavior and a depletion mode of MoS_2 transistor devices were observed. The MoS_2-TiO_2-Ti device with a TiO_2 interfacial layer showed more enhanced performance with a higher drive on current (I_{ON}). I_{ON} values for devices without and with the TiO_2 layer are 0.36 and 1.22 µA, respectively. The field effect mobility (μ_{FE}) values for MoS_2-Ti and MoS_2-TiO_2-Ti devices were estimated to be 1.38 and 6.08 cm²/V·s, respectively. The transfer curves at variable drain voltages and output characteristic also confirmed the better performance of the testing devices with the TiO_2 layer (Figure S2 in the Supplementary Materials information). The μ_{FE} values of the MoS_2-TiO_2-Ti device as a function of gate voltage were higher than those of the MoS_2-Ti device (Figure S3 in the Supplementary Materials information).

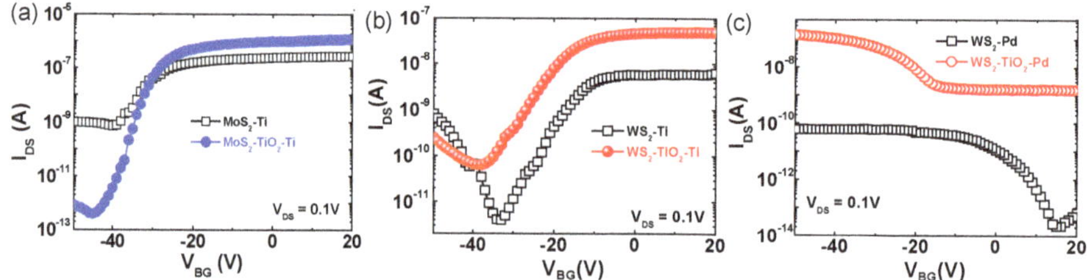

Figure 2. Transfer curves (I_{DS}-V_{BG}) for (**a**) MoS_2-Ti and MoS_2-TiO_2-Ti, (**b**) WS_2-Ti and WS_2-TiO_2-Ti, (**c**) and WS_2-Pd and WS_2-TiO_2-Pd.

A more interesting result was observed on the WS_2 FETs. Figure 2b shows a comparison of the I_{DS}-V_{BG} transfer characteristics of the WS_2-Ti and WS_2-TiO_2-Ti structured devices. The bi-polar behavior of the WS_2-Ti structured devices was observed, which is consistent with the previous results [34]. It is highly likely that the Fermi-level of the Ti metal exists within the mid-gap of WS_2. The transfer curve of the WS_2-TiO_2-Ti structured device showed stronger n-type unipolar behavior with a higher I_{ON} current than that of the WS_2-Ti device. As shown in Figure 2c, we also characterized the WS_2 devices using Pd metal electrodes with a relatively high work function of ~5.1 eV to understand the mid-gap pinning and the effects of an interfacial layer. The addition of the TiO_2 layer on the WS_2 caused a change from a weak bipolar to a p-type unipolar behavior. This result indicates that a high Schottky-barrier can be effectively reduced by a contact engineering approach utilizing a very thin TiO_2 interfacial layer. The I_{DS}-V_{BG} curves of the WS_2 FETs at various drain voltages are also shown in Figure S4 of the Supplementary Materials information. The performance enhancement of the 2D FET devices with the interfacial TiO_2 layer is attributed to the considerable reduction in the density of the diverse interface states, resulting from the direct contact between the metal and the 2D semiconductor channel. Comparison of the proposed band diagrams between the 2D TMD-Ti and 2D TMD-TiO_2-Ti devices highlights the change in the Schottky-barrier height as shown in Figure S5 in the Supplementary Materials information. In principle, the theoretical Fermi-level alignment between the metal and semiconductor, called Fermi-level depinning, also creates a more effective carrier modulation of the 2D TMD FET device.

For practical transistor applications, the electrical stability of the MoS_2 based FET devices was examined under a long-term positive gate-bias stress condition, as shown in Figure 3a–d. Figure 3a,b shows the shift of the I_{DS}-V_{BG} curves during the long-term gate-bias stress test. The transfer I-V curve properties were monitored every logarithmic time interval (1, 10, 100, 1000, and 10,000 s) while continuously applying +10 V to the gate electrode. Schemes to illustrate the stress measurement set up environment and the data checking points are shown in Figure S6 of the Supplementary Materials information. Even if the I_{DS}-V_{BG} curves in all of the MoS$_2$-Ti and MoS$_2$-TiO$_2$-Ti devices were slightly shifted to the positive direction, the device with the TiO$_2$ layer showed less of a shift than that without TiO$_2$, indicating more stable electrical properties compared to the control device without TiO$_2$. Interestingly, in Figure 3b, the variation of I_{OFF} values for the MoS$_2$-TiO$_2$-Ti stack seems more severe than that of the control MoS$_2$-Ti device. The actual differences of the minimum and maximum I_{OFF} values are 4.20×10^{-12} A and 3.68×10^{-10} A for MoS$_2$-Ti and MoS$_2$-TiO$_2$-Ti, respectively. The I_{OFF} fluctuation of all the devices was less than 1 nA, and this fluctuation was negligible in operation. Figure 3c shows a summary of the threshold voltage (V_{TH}) change for MoS$_2$-Ti and MoS$_2$-TiO$_2$-Ti stacked devices as a function of stress time, which was extracted from the raw data from Figure 3a,b. The MoS$_2$ FET without a TiO$_2$ layer showed a more positive V_{TH} shift than that of the MoS$_2$ FET with a TiO$_2$ layer. The V_{TH} shift for the MoS$_2$ FET without and with a TiO$_2$ interfacial layer was 3.1 and 1.1 V, respectively. The TiO$_2$ layer could serve as a buffer layer to mitigate the interfacial damage from electrical stress. As shown in Figure 3d, we also compared the field-effect mobility (μ_{FE}) values for the devices without and with a TiO$_2$ layer. The μ_{FE} was estimated by following equation:

$$\mu_{FE} = g_m \frac{L}{W} \frac{1}{V_{DS}} \frac{1}{C_{ox}} \text{ and } g_m = \frac{\partial I_D}{\partial V_G}$$

where g_m is the maximum transconductance that can be achieved from I_{DS}-V_{BG}, L is the channel length, W is the channel width, V_{DS} is the applied drain bias, and C_{ox} is the gate oxide capacitance.

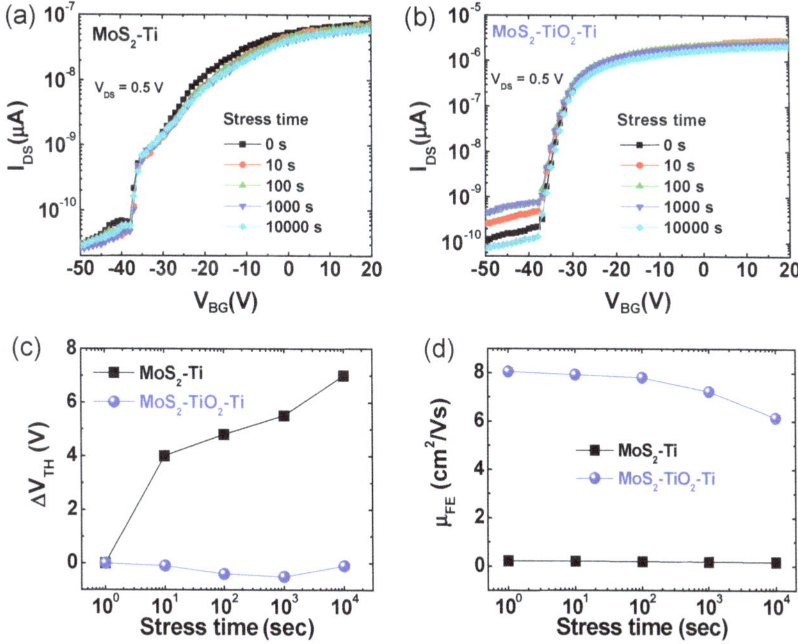

Figure 3. Transfer curves (I_{DS}-V_{BG}) of MoS$_2$ FETs (**a**) without TiO$_2$ and (**b**) with a TiO$_2$ layer during a 10,000 s gate-bias stress measurement at room temperature. The summary of (**c**) the ΔV_{TH} shift and (**d**) the μ_{FE} change as function of stress time for MoS$_2$-Ti and MoS$_2$-TiO$_2$-Ti.

Overall, the μ_{FE} of MoS_2-TiO_2-Ti device was higher than that of the MoS_2-Ti device. After 10,000 s stress time, the μ_{FE} was reduced from 0.22 to 0.17 cm^2/Vs for the device without a TiO_2 layer and from 8.05 to 6.14 cm^2/Vs for the device with a TiO_2 layer. Approximately, 25% of the μ_{FE} reduction was observed for both cases.

Additionally, the stability of the contact region for the WS_2-based FET devices was also determined for the effect of the interfacial TiO_2 layer on bias stress stability, as shown in Figure 4a,b. As can be seen, the transfer curves of the WS_2-Ti contact FET device showed bipolar behavior where both electron and hole carriers contribute to the current flow of the semiconductor channel. Overall, a lower V_{TH} shift was observed for the FET with a TiO_2 layer compared to the FET without a TiO_2 layer, indicating that the introduction of the TiO_2 interfacial layer on the WS_2 layered film is also an effective approach for improving the contact reliability of the WS_2 device, as well as the case of MoS_2 device. Specifically, the V_{TH} shifts for the WS_2 FET without and with a TiO_2 interfacial layer were 8 and 4.3 V, respectively (Figure 4c). As shown in Figure 4d, the change of μ_{FE} as a function of stress time was also fitted: the mobility value was almost unchanged for the control device without a TiO_2 layer and from 0.41 to 0.18 cm^2/Vs for the testing device with a TiO_2 layer.

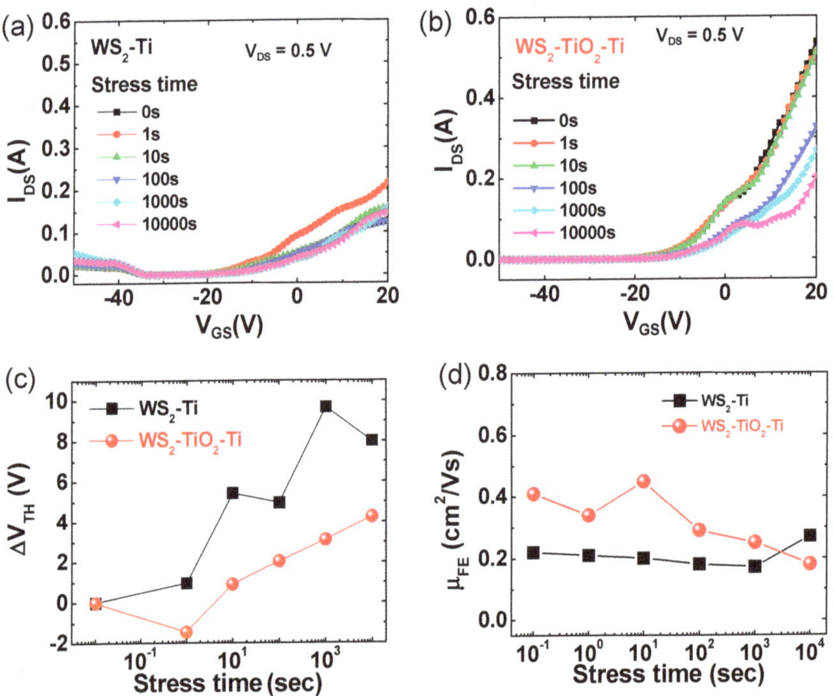

Figure 4. Transfer curves (I_{DS}-V_{BG}) of WS_2 FETs (**a**) without TiO_2 and (**b**) with a TiO_2 layer during a 10,000 s gate-bias stress measurement at room temperature. The summary of (**c**) the ΔV_{TH} shift and (**d**) the μ_{FE} change as a function of the stress time for WS_2-Ti and WS_2-TiO_2-Ti devices.

Indeed, the WS_2 FET device was more vulnerable to electrical stress than MoS_2, which might be due to greater number of interface states at the metal/semiconductor contact. The metal-induced gap states indispensably exist on the metal/semiconductor interface, which induces the instability of transistor performance. Additionally, there is a quantum mechanically long distance of 2–3 Å between the metal and 2D TMD, which increases the tunneling probability of charge carriers [35]. The more stable performance of the 2D TMD devices with an insulating TiO_2 layer might be understood by a mitigation of those gap states and a reduction in physical distance.

To unveil how the TiO_2 layer electronically influences the MoS_2 semiconductor, we explored a theoretical simulation of electronic states for free-standing MoS_2 and MoS_2/TiO_2 materials via a density functional theory (DFT) calculation (Figure 5). The density of states (DOS) calculation result of the free standing MoS_2 showed the existence of a forbidden gap (Figure 5a). Meanwhile, the TiO_2/MoS_2 hybrid combination featured a spin-polarized metallic behavior. The calculated DOS clearly validates that the addition of the TiO_2 layer leads to the modification of the electronical band structure of the junction region, offering the benefit of a doping effect on MoS_2.

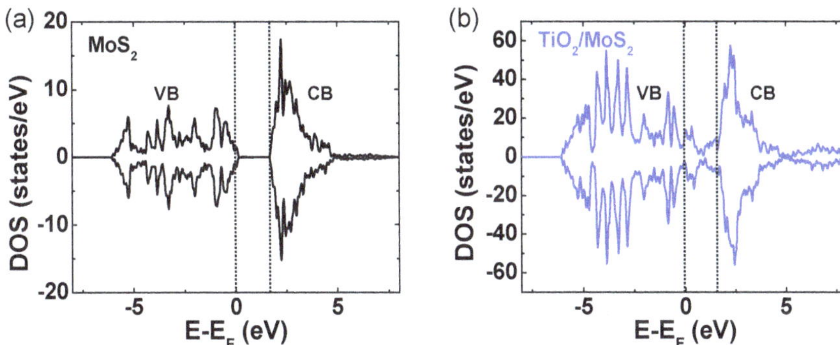

Figure 5. Density function theory (DFT)-calculated density of states (DOS) of (**a**) MoS_2 and (**b**) TiO_2/MoS_2.

4. Conclusions

The effect of a TiO_2 interfacial layer on metal/TMD (MoS_2 and WS_2) contact was experimentally and theoretically studied. The advantages of a Schottky-type FET device, possibly implemented according to the value of a metal work function, were achieved in the 2D TMD devices with a TiO_2 layer. Furthermore, a more enhanced and stable electrical performance for the 2D TMD FET devices with the TiO_2 interfacial layer could be obtained under a gate-bias stress condition. The TiO_2 interfacial layer could serve as a Fermi-level de-pinning layer, reducing the density of the interface states. Additionally, the DFT calculation validates the doping effect of the TiO_2 interfacial layer on the 2D MoS_2. The strategy of inserting a very thin insulating layer into the contact region will be also applied to diverse 2D TMD-based FET devices.

Supplementary Materials: The following are available online at http://www.mdpi.com/2079-4991/9/8/1155/s1, Figure S1: Fabrication process and band diagram for Fermi-level pinning with various metals; Figure S2: Transfer and output curves for MoS_2 FETs without and with a TiO_2 layer; Figure S3: Field-effect mobility (μFE) for MoS_2 FETs without and with a TiO_2 layer; Figure S4: Transfer curves for WS_2-Ti, WS_2-TiO_2-Ti, WS_2-Pd, and WS_2-TiO_2-Pd structured FET devices; Figure S5: Energy band diagrams for TMDC-Ti and TMDC-TiO_2-Ti stacks; Figure S6: Bias stress measurement sequence.

Author Contributions: W.P. designed and conducted the experiments, and H.Y.J., J.H.N., and S.O. supported the electrical measurement and analysis. T.H.K. set up the experiment system. Y.P., Y.K. and S.M.C. supported the process of experiments and the analysis of data. B.C. supported and guided the experiment and the results. B.C. conceived and advised the publication of paper.

Funding: This work was supported by the National Research Foundation of Korea (NRF) grant funded by the Korean government (MSIT; Ministry of Science and ICT) (No. 2017R1C1B1005076) and Fundamental Research Program (No. PNK6300) of the Korea Institute of Materials Science (KIMS). This research was also financially supported by the Ministry of Trade, Industry and Energy (MOTIE) and Korea Institute for Advancement of Technology (KIAT) through the National Innovation Cluster R & D program (P0006704_Development of energy saving advanced parts).

Conflicts of Interest: The authors declare no conflict of interest.

References

1. Radisavljevic, B.; Radenovic, A.; Brivio, J.; Giacometti, V.; Kis, A. Single-layer MoS$_2$ transistors. *Nat. Nanotechnol.* **2011**, *6*, 147–150. [CrossRef] [PubMed]

2. Bertolazzi, S.; Brivio, J.; Kis, A. Stretching and Breaking of Ultrathin MoS$_2$. *ACS Nano* **2011**, *5*, 9703–9709. [CrossRef] [PubMed]

3. Yu, W.J.; Li, Z.; Zhou, H.; Chen, Y.; Wang, Y.; Huang, Y.; Duan, X. Vertically stacked multi-heterostructures of layered materials for logic transistors and complementary inverters. *Nat. Mater.* **2013**, *12*, 246–252. [CrossRef] [PubMed]

4. Fang, H.; Tosun, M.; Seol, G.; Chang, T.C.; Takei, K.; Guo, J.; Javey, A. Degenerate n-Doping of Few-Layer Transition Metal Dichalcogenides by Potassium. *Nano Lett.* **2013**, *13*, 1991–1995. [CrossRef] [PubMed]

5. Castellanos-Gomez, A.; Poot, M.; Steele, G.A.; van der Zant, H.S.J.; Agraït, N.; Rubio-Bollinger, G. Elastic Properties of Freely Suspended MoS2 Nanosheets. *Adv. Mater.* **2012**, *24*, 772–775. [CrossRef]

6. Salvatore, G.A.; Münzenrieder, N.; Barraud, C.; Petti, L.; Zysset, C.; Büthe, L.; Ensslin, K.; Tröster, G. Fabrication and Transfer of Flexible Few-Layers MoS2 Thin Film Transistors to Any Arbitrary Substrate. *ACS Nano* **2013**, *7*, 8809–8815. [CrossRef]

7. Li, H.; Yin, Z.; He, Q.; Li, H.; Huang, X.; Lu, G.; Fam, D.W.H.; Tok, A.I.Y.; Zhang, Q.; Zhang, H. Fabrication of Single- and Multilayer MoS$_2$ Film-Based Field-Effect Transistors for Sensing NO at Room Temperature. *Small* **2012**, *8*, 63–67. [CrossRef]

8. Radisavljevic, B.; Whitwick, M.B.; Kis, A. Integrated Circuits and Logic Operations Based on Single-Layer MoS$_2$. *ACS Nano* **2011**, *5*, 9934–9938. [CrossRef]

9. Li, H.; Lu, G.; Yin, Z.; He, Q.; Li, H.; Zhang, Q.; Zhang, H. Optical Identification of Single- and Few-Layer MoS$_2$ Sheets. *Small* **2012**, *8*, 682–686. [CrossRef]

10. Roy, T.; Tosun, M.; Cao, X.; Fang, H.; Lien, D.-H.; Zhao, P.; Chen, Y.-Z.; Chueh, Y.-L.; Guo, J.; Javey, A. Dual-Gated MoS$_2$ /WSe$_2$ van der Waals Tunnel Diodes and Transistors. *ACS Nano* **2015**, *9*, 2071–2079. [CrossRef]

11. Wi, S.; Kim, H.; Chen, M.; Nam, H.; Guo, L.J.; Meyhofer, E.; Liang, X. Enhancement of Photovoltaic Response in Multilayer MoS$_2$ Induced by Plasma Doping. *ACS Nano* **2014**, *8*, 5270–5281. [CrossRef] [PubMed]

12. Kim, S.; Konar, A.; Hwang, W.-S.; Lee, J.H.; Lee, J.; Yang, J.; Jung, C.; Kim, H.; Yoo, J.-B.; Choi, J.-Y.; et al. High-mobility and low-power thin-film transistors based on multilayer MoS$_2$ crystals. *Nat. Commun.* **2012**, *3*, 1011. [CrossRef] [PubMed]

13. Wang, H.; Yu, L.; Lee, Y.-H.; Shi, Y.; Hsu, A.; Chin, M.L.; Li, L.-J.; Dubey, M.; Kong, J.; Palacios, T. Integrated Circuits Based on Bilayer MoS$_2$ Transistors. *Nano Lett.* **2012**, *12*, 4674–4680. [CrossRef] [PubMed]

14. Jo, S.; Ubrig, N.; Berger, H.; Kuzmenko, A.B.; Morpurgo, A.F. Mono- and Bilayer WS2 Light-Emitting Transistors. *Nano Lett.* **2014**, *14*, 2019–2025. [CrossRef] [PubMed]

15. Lee, H.S.; Min, S.-W.; Park, M.K.; Lee, Y.T.; Jeon, P.J.; Kim, J.H.; Ryu, S.; Im, S. MoS2 Nanosheets for Top-Gate Nonvolatile Memory Transistor Channel. *Small* **2012**, *8*, 3111–3115. [CrossRef] [PubMed]

16. Kwon, J.; Ki Hong, Y.; Kwon, H.-J.; Jin Park, Y.; Yoo, B.; Kim, J.; Grigoropoulos, C.P.; Suk Oh, M.; Kim, S. Optically transparent thin-film transistors based on 2D multilayer MoS$_2$ and indium zinc oxide electrodes. *Nanotechnology* **2015**, *26*, 035202. [CrossRef] [PubMed]

17. Zhang, W.; Chiu, M.-H.; Chen, C.-H.; Chen, W.; Li, L.-J.; Wee, A.T.S. Role of Metal Contacts in High-Performance Phototransistors Based on WSe$_2$ Monolayers. *ACS Nano* **2014**, *8*, 8653–8661. [CrossRef] [PubMed]

18. Yin, Z.; Li, H.; Li, H.; Jiang, L.; Shi, Y.; Sun, Y.; Lu, G.; Zhang, Q.; Chen, X.; Zhang, H. Single-Layer MoS$_2$ Phototransistors. *ACS Nano* **2012**, *6*, 74–80. [CrossRef] [PubMed]

19. Park, J.; Kang, D.-H.; Kim, J.-K.; Park, J.-H.; Yu, H.-Y. Efficient Threshold Voltage Adjustment Technique by Dielectric Capping Effect on MoS$_2$ Field-Effect Transistor. *IEEE Electron Device Lett.* **2017**, *38*, 1172–1175. [CrossRef]

20. Zheng, H.M.; Gao, J.; Sun, S.M.; Ma, Q.; Wang, Y.P.; Zhu, B.; Liu, W.J.; Lu, H.L.; Ding, S.J.; Zhang, D.W. Effects of Al$_2$O$_3$ Capping and Post-Annealing on the Conduction Behavior in Few-Layer Black Phosphorus Field-Effect Transistors. *IEEE J. Electron Devices Soc.* **2018**, *6*, 320–324. [CrossRef]

21. Agrawal, A.; Lin, J.; Zheng, B.; Sharma, S.; Chopra, S.; Wang, K.; Gelatos, A.; Mohney, S.; Datta, S. Barrier Height Reduction to 0.15eV and Contact Resistivity Reduction to 9.1×10^{-9} Ω-cm^2 Using Ultrathin TiO2-x Interlayer between Metal and Silicon. In Proceedings of the 2013 Symposium on VLSI Technology, Kyoto, Japan, 11–14 June 2013.

22. Yang, L.; Majumdar, K.; Liu, H.; Du, Y.; Wu, H.; Hatzistergos, M.; Hung, P.Y.; Tieckelmann, R.; Tsai, W.; Hobbs, C.; et al. Chloride Molecular Doping Technique on 2D Materials: WS_2 and MoS_2. *Nano Lett.* **2014**, *14*, 6275–6280. [CrossRef] [PubMed]

23. Park, W.; Kim, Y.; Jung, U.; Yang, J.H.; Cho, C.; Kim, Y.J.; Hasan, S.M.N.; Kim, H.G.; Lee, H.B.R.; Lee, B.H. Complementary Unipolar WS_2 Field-Effect Transistors Using Fermi-Level Depinning Layers. *Adv. Electron. Mater.* **2016**, *2*, 1500278. [CrossRef]

24. Park, W.; Kim, Y.; Lee, S.K.; Jung, U.; Yang, J.H.; Cho, C.; Kim, Y.J.; Lim, S.K.; Hwang, I.S.; Lee, B.H. Contact resistance reduction using Fermi level de-pinning layer for MoS2 FETs. In Proceedings of the 2014 IEEE International Electron Devices Meeting, San Francisco, CA, USA, 15–17 December 2014; 2014; pp. 1–4.

25. Agrawal, A.; Lin, J.; Barth, M.; White, R.; Zheng, B.; Chopra, S.; Gupta, S.; Wang, K.; Gelatos, J.; Mohney, S.E.; et al. Fermi level depinning and contact resistivity reduction using a reduced titania interlayer in n-silicon metal-insulator-semiconductor ohmic contacts. *Appl. Phys. Lett.* **2014**, *104*, 112101. [CrossRef]

26. Das, S.; Chen, H.-Y.; Penumatcha, A.V.; Appenzeller, J. High Performance Multilayer MoS_2 Transistors with Scandium Contacts. *Nano Lett.* **2013**, *13*, 100–105. [CrossRef]

27. Noori, A.M.; Balseanu, M.; Boelen, P.; Cockburn, A.; Demuynck, S.; Felch, S.; Gandikota, S.; Gelatos, A.J.; Khandelwal, A.; Kittl, J.A.; et al. Manufacturable Processes for $\leq$ 32-nm-node CMOS Enhancement by Synchronous Optimization of Strain-Engineered Channel and External Parasitic Resistances. *IEEE Trans. Electron Devices* **2008**, *55*, 1259–1264. [CrossRef]

28. Du, Y.; Liu, H.; Neal, A.T.; Si, M.; Ye, P.D. Molecular Doping of Multilayer MoS_2 Field-Effect Transistors: Reduction in Sheet and Contact Resistances. *IEEE Electron Device Lett.* **2013**, *34*, 1328–1330. [CrossRef]

29. Park, W.; Min, J.-W.; Shaikh, S.F.; Hussain, M.M. Stable MoS_2 Field-Effect Transistors Using TiO_2 Interfacial Layer at Metal/MoS_2 Contact. *Phys. Status Solidi A* **2017**, *214*, 1700534. [CrossRef]

30. Late, D.J.; Liu, B.; Matte, H.S.S.R.; Dravid, V.P.; Rao, C.N.R. Hysteresis in Single-Layer MoS_2 Field Effect Transistors. *ACS Nano* **2012**, *6*, 5635–5641. [CrossRef] [PubMed]

31. Addou, R.; McDonnell, S.; Barrera, D.; Guo, Z.; Azcatl, A.; Wang, J.; Zhu, H.; Hinkle, C.L.; Quevedo-Lopez, M.; Alshareef, H.N.; et al. Impurities and Electronic Property Variations of Natural MoS_2 Crystal Surfaces. *ACS Nano* **2015**, *9*, 9124–9133. [CrossRef] [PubMed]

32. Cho, A.-J.; Yang, S.; Park, K.; Namgung, S.D.; Kim, H.; Kwon, J.-Y. Multi-Layer MoS_2 FET with Small Hysteresis by Using Atomic Layer Deposition Al_2O_3 as Gate Insulator. *ECS Solid State Lett.* **2014**, *3*, Q67–Q69. [CrossRef]

33. Rehman, A.U.; Khan, M.F.; Shehzad, M.A.; Hussain, S.; Bhopal, M.F.; Lee, S.H.; Eom, J.; Seo, Y.; Jung, J.; Lee, S.H. n-MoS2 /p-Si Solar Cells with Al_2O_3 Passivation for Enhanced Photogeneration. *ACS Appl. Mater. Interfaces* **2016**, *8*, 29383–29390. [CrossRef] [PubMed]

34. Sik Hwang, W.; Remskar, M.; Yan, R.; Protasenko, V.; Tahy, K.; Doo Chae, S.; Zhao, P.; Konar, A.; (Grace) Xing, H.; Seabaugh, A.; et al. Transistors with chemically synthesized layered semiconductor WS_2 exhibiting 105 room temperature modulation and ambipolar behavior. *Appl. Phys. Lett.* **2012**, *101*, 013107. [CrossRef]

35. Kang, J.; Liu, W.; Banerjee, K. High-performance MoS_2 transistors with low-resistance molybdenum contacts. *Appl. Phys. Lett.* **2014**, *104*, 093106. [CrossRef]

 *nanomaterials*

Article

Preparation and Tribological Properties of WS$_2$ Hexagonal Nanoplates and Nanoflowers

Xianghua Zhang [1,*], Jiangtao Wang [2], Hongxiang Xu [1], Heng Tan [1] and Xia Ye [1]

[1]   School of Mechanical Engineering, Jiangsu University of Technology, Changzhou 213001, China; jxxhx@jsut.edu.cn (H.X.); tanheng@jsut.edu.cn (H.T.); yexia@jsut.edu.cn (X.Y.)

[2]   School of Materials Science and Engineering, Jiangsu University of Technology, Changzhou 213001, China; jxwjt@jsut.edu.cn

*   Correspondence: zxh@jsut.edu.cn; Tel.: +86-519-8695-3212

Received: 27 April 2019; Accepted: 27 May 2019; Published: 1 June 2019

Abstract: This paper presents the facile synthesis of two different morphologies of WS$_2$ nanomaterials—WS$_2$ hexagonal nanoplates and nanoflowers—by a sulfurization reaction. The phases and morphology of the samples were investigated by X-ray diffraction (XRD), scanning electron microscopy (SEM), and transmission electron microscopy (TEM). The tribological performance of the two kinds of WS$_2$ nanomaterials as additives in paraffin oil were measured using a UMT (Universal Mechanical Tester)-2 tribotester. The results demonstrated that the friction and wear performance of paraffin oil can be greatly improved with the addition of WS$_2$ nanomaterials, and that the morphology and content of WS$_2$ nanomaterials have a significant effect on the tribological properties of paraffin oil. The tribological performance of lubricating oil was best when the concentration of the WS$_2$ nanomaterial additive was 0.5 wt %. Moreover, the paraffin oil with added WS$_2$ nanoflowers exhibited better tribological properties than paraffin oil with added WS$_2$ hexagonal nanoplates. The superior tribological properties of WS$_2$ nanoflowers can be attributed to their special morphology, which contributes to the formation of a uniform tribo-film during the sliding process.

Keywords: WS$_2$; lubricant additives; tribological properties

1. Introduction

In recent years, the global energy crisis and environmental pollution have been serious problems. Now, the regulatory requirements for reducing energy consumption and avoiding energy losses are becoming more stringent. Because of this, reducing energy consumption and greenhouse gas emissions has become an important focus for researchers. According to recent research by Holmberg et al., the friction of engines, gearboxes, tires, auxiliary equipment, and brakes in heavy vehicles consumes 33% of fuel energy [1], friction in cars consumes 28% of fuel energy [2], and the energy consumed by internal friction in the entire paper mill accounts for 15–25% [3]. Therefore, many attempts have been made to introduce various methods to overcome friction. Lubrication is known to be one of the most effective ways to reduce friction and wear, and the antifriction effect of lubricating oil is mainly affected by the lubricant additive. Recent studies have found that some nanomaterials have good antifriction performance due to their special structure. Therefore, increasing attention is now being paid to the use of nanomaterials as lubricant additives to improve the tribological properties of lubricating oil.

In the past few years, a variety of nanomaterials have been used as lubricant additives, and their tribological properties have been extensively studied. These materials can be classified into the following categories. The first type is metallic nanoparticles, including Cu, Fe, Ni, etc. [4–6]. The second includes carbon materials such as carbon nanotubes and graphene [7–10]. The third is composed of the transition metal chalcogenides, containing MoS$_2$, WS$_2$, MoSe$_2$, WSe$_2$, etc. [11–15]. The last category comprises other nanomaterials such as oxides, fluorides, and borides [16–20]. Among these different types of materials, transition metal chalcogenides have received great attention due to their special layered structure.

WS$_2$, as an important member of the transition metal chalcogenide material family, has attracted great attention for its intriguing electronic, electrochemical, and electrocatalytic properties, and for its extensive applications in field-effect transistors, energy storage, catalysis, and hydrogen storage media [21–28]. In addition, WS$_2$ is an excellent solid lubricant due to its special layered structure, which is composed of strong S–W–S covalent bonds inside the layers, and the weak van der Waals force between the layers. The easy sliding between WS$_2$ layers under small shear forces is often regarded as an important feature of its excellent lubricity [29]. Recently, WS$_2$ nanomaterials with different morphologies have been synthesized, and their tribological properties and antifriction mechanisms have been studied. For example, Tenne et al. [11] investigated the tribological properties of fullerene-like WS$_2$ nanoparticles as additives in a lubricating oil under harsh conditions, and the results showed that WS$_2$ nanoparticles play a major role in alleviating friction and wear. Wu et al. [30] synthesized hollow WS$_2$ spheres by a solvothermal process and compared their tribological properties with commercial colloidal MoS$_2$ as an additive in liquid paraffin. Zhang et al. [12] prepared WS$_2$ nanorods by using a self-transformation process and investigated the tribological performance of WS$_2$ nanorods as an additive in lubricating oil. It was found that the antiwear ability of the base oil was improved by the addition of WS$_2$ nanorods. Hu et al. [31] studied the tribological properties of WS$_2$ and WS$_2$/TiO$_2$ nanoparticles dispersed in diisooctyl sebacate and found that the two nano-additives slightly affected the friction reduction effect, but WS$_2$/TiO$_2$ nanoparticles were found to remarkably improve the wear resistance of diisooctyl sebacate.

All of the above studies have shown that WS$_2$ nanomaterials with different morphologies help to improve the tribological properties of lubricating oils. However, these studies have only investigated the tribological properties of WS$_2$ nanomaterials with a single morphology, and did not explore the antifriction properties and mechanisms of WS$_2$ nanomaterials with a different morphology under the same working conditions. Previous studies investigating MoS$_2$ nanomaterials have demonstrated the complex relationship between the morphology size and tribological properties of MoS$_2$. For example, Xu et al. reported that the lubricity of the sheet-like nano-MoS$_2$ is inferior to that of the micro-scale MoS$_2$ in rapeseed oil [32]. However, Raboso et al. reported that the size and morphology of MoS$_2$ did not have a significant effect on the friction and wear of the polyalphaolefin oil [33]. Therefore, it is valuable to study the tribological properties and friction reduction mechanism of WS$_2$ nanomaterials with different morphologies.

In this study, two different morphologies of WS$_2$ nanomaterials—WS$_2$ hexagonal nanoplates and nanoflowers—were synthesized by a different high-temperature solid-phase reaction process. The tribological properties of the two kinds of WS$_2$ nanomaterials as additives in the paraffin oil were also investigated.

2. Materials and Methods

2.1. Reagents and Materials

Tungsten and sulfur powders were purchased from Sinopharm Chemical Reagent Co. Ltd (Shanghai, China). Tungsten trioxide and thiourea were obtained from the Aladdin Chemical Reagent Company (Shanghai, China). All chemical reagents were used directly without further purification.

2.2. Synthesis of WS$_2$ Hexagonal Nanoplates

In a typical method, high-purity tungsten and sulfur powder (W:S molar ratio of 1:3, S powder excess 50%) were poured into a steel kettle, and then the powders were mechanically ground with a speed of 300 rpm (rotations per minute) in a planetary ball mill for 12 h. Then, the ball-milled mixture was transferred into a stainless-steel reactor. The reactor was tightly closed and pushed into the middle of a tube furnace. The temperature of the tube furnace was raised to 650 °C at a rate of 5 °C min^{-1} in an atmosphere of N_2 and the temperature was maintained at 650 °C for 2 h. Subsequently, the tube was gradually cooled to room temperature and the prepared powders were then obtained.

2.3. Synthesis of WS$_2$ Nanoflowers

The WS$_2$ nanoflowers were synthesized according to the previous method [34] reported by us with minor modifications. This included 10 mmol of WO_3, 60 mmol of sulfur powder and 140 mmol of thiourea were ground in a mortar for 30 min. Then 3 g of the ground mixture was loaded in an alumina boat. This boat was pushed into the hot zone of the tube furnace. The furnace temperature was maintained at 850 °C for 1 h in N_2 atmosphere and then gradually cooled to room temperature.

2.4. Materials Characterization

The X-ray diffraction (XRD) pattern was recorded by a Shimadzu LabX XRD-6000 X-ray diffractometer using a Cu Ka X-ray source operating at 40 kV and 30 mA with a scanning range of 10° to 80°. A JSM-7001F field-emission scanning electron microscope (FESEM) and a JEM-2100 transmission electron microscope (TEM) were used to record the sample morphology.

2.5. Tribological Properties Test

A UMT-2 tribotester (CETR, San Jose, CA, USA) was used to measure the tribological properties of the two WS$_2$ samples. The prepared WS$_2$ powders were dispersed into the paraffin oil by ultrasonic dispersion for 60 min which then resulted in the required lubricating oil with different WS$_2$ contents. The tribological properties tests were performed in ball-and-disk mode with a load of 10–60 N and a rotational speed of 100–400 rpm for 30 min. The friction pair consisted of a ball with a diameter of 10 mm and a disc with size of $\Phi 40$ mm × 3 mm. The fixed upper sample (ball) is made of GCr15 bearing steel (AISI 52100) with a hardness of 62 HRC (Rockwellhardness) and the rotating lower sample (disk) is made of 45# steel. The surface of the steel disc was polished and cleaned with acetone before the test. The friction coefficient was automatically recorded during the contact friction, and the widths of the wear scars were measured by an optical microscope. The morphologies and elements of the wear scars on the surface of the lower disc were investigated by scanning electron microscopy (SEM) and energy-dispersive X-ray spectroscopy (EDS).

3. Results and Discussion

3.1. Structure and Morphology Characterization

The crystal structure and phase purity of the synthesized samples were verified by the XRD patterns, as presented in Figure 1. From the image, it can be seen that the diffraction patterns of the two samples were significantly different. The diffraction peaks of the nanoplates located at 14.30°, 28.84°, 32.74°, 33.50°, 39.50°, 43.90°, 49.70°, 58.40°, 59.80°, 60.48°, and 66.50° were assigned to the (002), (004), (100), (101), (103), (006), (105), (110), (008), (112), and (114) planes of WS$_2$, respectively. Furthermore, a high and sharp (002) peak was observed from the XRD pattern, indicating that the WS$_2$ nanoplates were stacked together with a highly ordered packing [35]. In contrast, only (002), (100), (101), and (110) peaks could be detected in the diffraction pattern of the nanoflower sample. Besides, the intensity of the (002) peak located at 13.76° was significantly weakened, and its position was shifted to the left by 0.56° from the standard card. This indicates that the number of stacks in the (002) layer of the WS$_2$ nanoflowers was reduced, and the layer interval became larger [36]. All the diffraction peaks of the two patterns could be indexed to the hexagonal phase (p63/mmc space group) of WS$_2$ (JSPDS No. 08-0237). No evidence of any other phases was detected, indicating that the samples were of high purity.

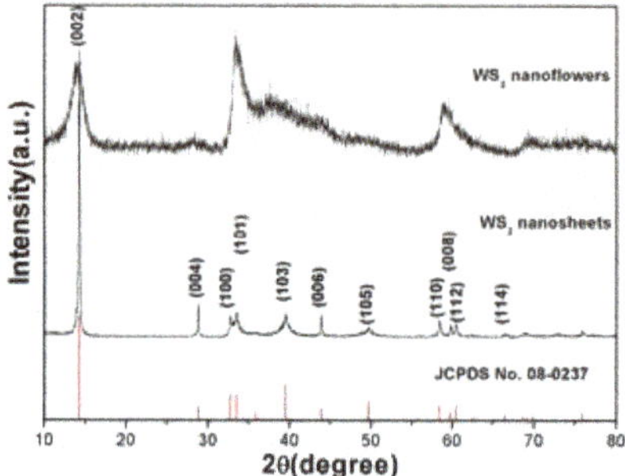

Figure 1. X-ray diffraction (XRD) pattern of the as-synthesized WS_2 hexagonal nanoplates and nanoflowers.

The morphology and size of the two fabricated WS_2 samples were identified by SEM and TEM. The SEM images of the WS_2 hexagonal nanoplates are presented in Figure 2a,b. From the low-magnification SEM image (Figure 2a), it can be seen that the sample was composed of a large number of regular nanoplates with the diameter of about 0.5–1 µm. The SEM image with higher magnification in Figure 2b presents a clear view of the surface morphology of the nanoplates. These nanoplates exhibited hexagonal morphology with a thickness of 50–100 nm. Figure 2c,d displays the SEM images of the WS_2 nanoflowers. Some agglomerated WS_2 flower-like structures are presented in Figure 2c. It can be seen from the enlarged image (Figure 2d) that these nanoflowers were composed of some ultrathin nanosheets, and the edges of these nanosheets were obviously curled. That is because these nanosheets are unstable and tend to form a closed structure by rolling up, thereby reducing the number of dangling bonds and the total energy of the system [37].

To further reveal the morphology and microstructure of these WS_2 nanomaterials, TEM measurements were performed on the samples. As shown in Figure 2e, perfectly hexagonal WS_2 nanoplates with diameters of 3.5 µm were observed. The TEM image of the WS_2 nanoflower is shown in Figure 2f, from which it can be seen that the WS_2 nanoflowers were dispersed into ultrathin nanosheets after sonication, but the nanosheets were still connected together. In addition, the edges of the nanosheets were significantly curled, which is consistent with the SEM photographs.

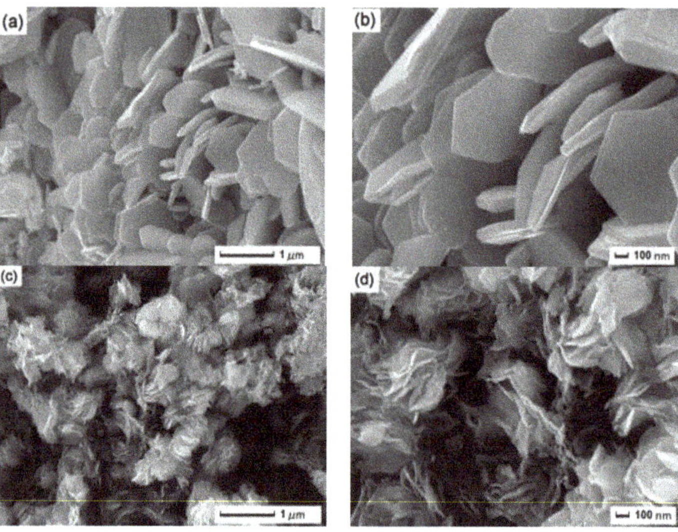

Figure 2. *Cont.*

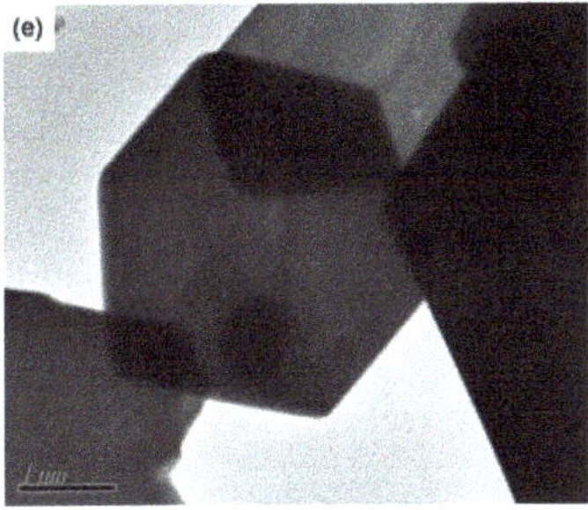

Figure 2. Scanning electron microscopy (SEM) images of WS$_2$ hexagonal nanoplates (**a,b**) and nanoflowers (**c,d**); transmission electron microscopy (TEM) images of WS$_2$ hexagonal nanoplates (**e**) and nanoflowers (**f**).

3.2. Analysis of Tribological Properties

The tribological properties of the two different WS$_2$ nanomaterials as lubricant additives in paraffin oil were investigated by a UMT-2 tribotester. Figure 3a shows the effect of the nanomaterial additive concentrations on the tribological properties with a working load of 20 N at 200 rpm for 30 min. From this, it could be found that the average friction coefficient of paraffin oil with the addition of the two different nanomaterials was smaller than that of pure paraffin oil. When the content of the additives was 0.5 wt %, the friction coefficient of the paraffin oil with WS$_2$ nanoflowers reached the lowest value, a reduction of 29.1% in comparison with pure paraffin oil, while that of paraffin oil containing WS$_2$ nanoplates was only reduced by 24.5%. Additionally, when the concentration of the nanomaterials was higher than 0.5 wt %, the friction reducing performance was gradually weakened with the increase of the additive concentration. It can be concluded that the friction coefficient will increase when the additive content is too low or too high. The reason is that when the concentration is too low, a continuous lubricating film cannot be formed on the surface of the friction pair, and when the concentration is too high, the additive will agglomerate, which affects the friction-reducing effect [38]. Figure 3b exhibits the real-time friction coefficient curve of pure paraffin oil and the two lubricating oils with 0.5 wt % nanomaterial added. In the beginning, the three curves had the same trend, and the friction coefficient changed from large to small, which is attributable to the lack of lubricant between the friction pairs. With the embedding of the lubricant, the friction coefficient was drastically reduced. However, after 10 min, the friction coefficient of the pure paraffin oil began to increase, and the coefficient fluctuated greatly. However, the friction coefficient of the paraffin oil with added nanomaterials was very stable, and the friction coefficient of the WS$_2$ nanoflowers was always lower than that of the nanoplates. The above experimental results indicate that the paraffin oil containing WS$_2$ nanoflowers possessed better lubricating properties than both the pure paraffin oil and the paraffin oil containing WS$_2$ nanoplates.

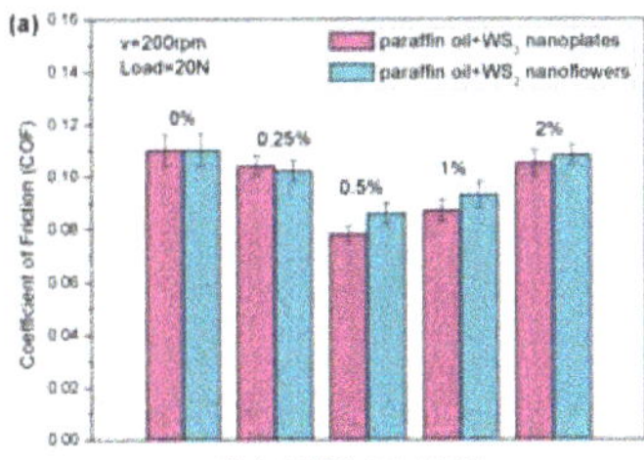

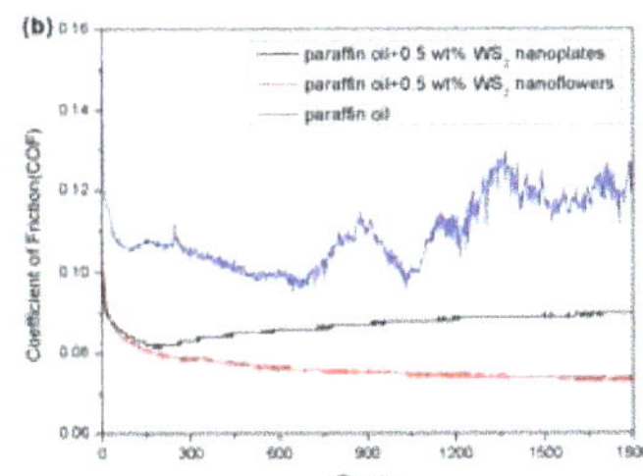

Figure 3. (**a**) The changes of the average friction coefficient of the WS$_2$ nanoplates and nanoflowers with different concentration, and (**b**) the real-time friction coefficient as a function of sliding time when lubricated by three different oil samples.

In order to further compare the tribological properties of the two WS_2 nanomaterials, comparative experiments were carried out with different loads and different rotating speeds. Figure 4a shows the average friction coefficient as a function of applied load when the additive concentration was 0.5 wt % and the tribotester was operated with a rotating speed of 200 rpm for 30 min. Obviously, when the applied load was increased from 10 to 40 N, the average friction coefficient had a downward trend. However, when the load was increased to 60 N, the coefficient increased slightly. The variation of the average friction coefficient of the two kinds of WS_2 nanomaterials with the change of rotating speed is presented in Figure 4b. The trend of the average friction coefficient in Figure 4b is similar to that in Figure 4a, which indicates that the friction coefficient first decreased with increasing speed, and then increased. In addition, under the same conditions, the average friction coefficient of the WS_2 nanoflowers was always lower than that of the WS_2 nanoplates.

From the above results, it was found that WS_2 nanoflowers exhibited better tribological properties, and it was found that the friction coefficient could be remarkably decreased by adding two kinds of WS_2 nanomaterials into the paraffin oil.

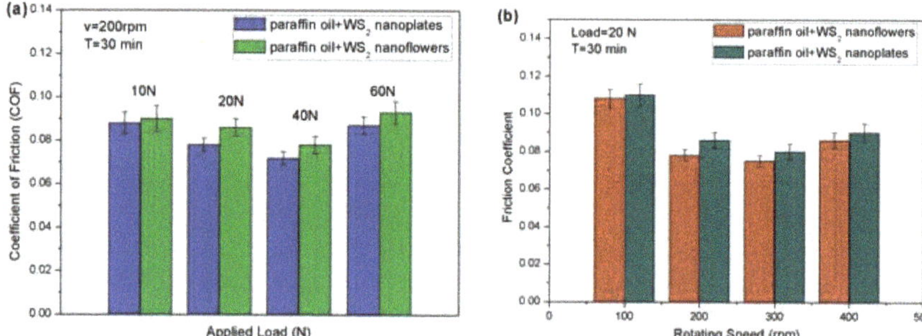

Figure 4. The changes of average coefficient of friction of the WS_2 nanoflowers and nanoplates with different load (**a**) and different rotating speed (**b**).

In order to compare and analyze the antiwear properties of the two kinds of WS_2 nanomaterials, the wear surface of the steel disc was examined by optical microscopy and SEM. The optical micrograph and SEM images of the wear surface lubricated with pure paraffin oil, and paraffin oil with 0.5 wt % added WS_2 nanoplates and nanoflowers are shown in Figure 5. The test time was 30 min and the test load was 20 N. It can be seen from Figure 5a that when the friction pair was lubricated with the pure paraffin oil, the width of the wear scar was about 453 μm. When WS_2 nanoplates and nanoflowers were added into the paraffin oil, the width of the wear scar was significantly reduced to 373 μm (Figure 5c) and 340 μm (Figure 5e). Furthermore, some deep and wide furrows were observed on the wear surface shown in Figure 5a, which clearly indicates that the surface was subjected to a large contact stress during sliding. The same result was also found in the SEM image. As shown in Figure 5b, many grooves and pits were discovered on the surface of the wear scar, and some abrasive grains with different sizes were attached to it. When 0.5 wt.% of WS_2 nanoplates were added into the paraffin oil, the surface topography of the wear scars was significantly improved. A dark-colored tribo-film can be observed in the optical picture (Figure 5c), but the tribo-film was unevenly distributed. In the TEM image (Figure 5d), only some very shallow grooves and some very small grinding debris can be found. Compared to the above two lubricants, the friction surface added with the WS_2 nanoflower lubricant was the least damaged. A large number of dark areas (tribo-film) were observed on the track in Figure 5e and only a few very shallow grooves were seen in Figure 5f and no wear debris was found. These results verify that WS_2 nanosheets and nanoflowers can improve the antiwear performance of paraffin oil, but the antiwear ability of WS_2 nanoflowers is better than that of nanoplates.

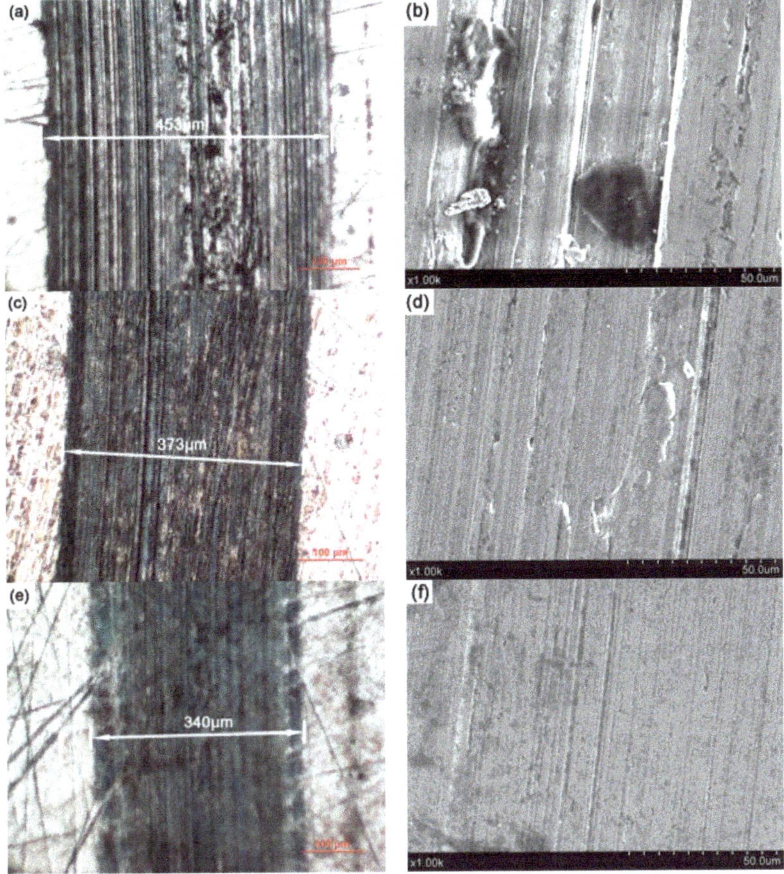

Figure 5. Optical images and SEM micrographs of wear scars lubricated with pure paraffin oil (**a,b**), paraffin oil + 0.5 wt % WS$_2$ nanoplates (**c,d**), and paraffin oil + WS$_2$ nanoflowers (**e,f**).

In order to investigate the lubrication mechanism of the WS$_2$ nanoplates and nanoflowers, EDS was used to investigate the worn surface. The EDS spectra obtained from the worn scar of Figure 5d,f are presented in Figure 6. The elements of W and S were present on the worn surface. This could prove that there was WS$_2$ deposited on the worn surface during the process of friction.

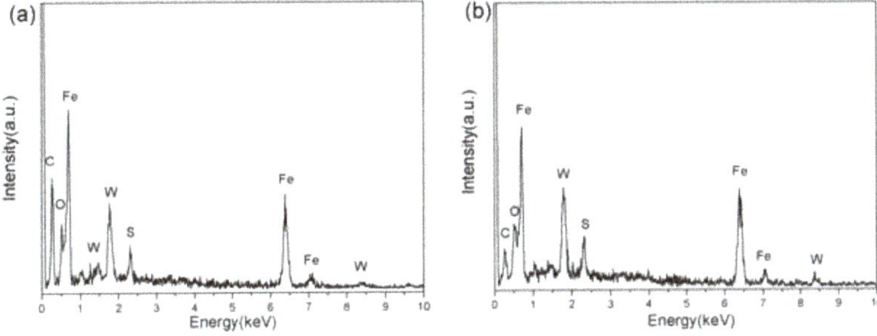

Figure 6. Energy-dispersive X-ray spectroscopy (EDS) of the worn scar of a steel disc lubricated with 0.5 wt % WS$_2$ nanoplates (**a**) and nanoflowers (**b**).

There is a great deal of literature regarding the antifriction and antiwear mechanisms of nanomaterials as lubricant additives, and they can be summarized into the following three reasons. The first is that the nanomaterial produces a rolling effect on the surface of the friction pair [12]. The second reason is that the nanomaterial adsorbed on the surface of the friction pair forms a lubricating film [13]. The last reason is that nanomaterials have a repair effect on the surface of the friction pair [4,6].

According to the above experimental results, we could infer the reasons for the friction reduction and antiwear properties of the WS_2 nanoplates and nanoflowers. The main reason can be attributed to the formation of a tribo-film on the rubbing surface, but there was still a difference in the mechanism of the antifriction and antiwear between the WS_2 nanoplate and the nanoflower at the beginning. When the WS_2 nanoplates were used as lubricant additive, the WS_2 nanoplates would penetrate into the interface of the friction surface. However, due to the large thickness of the nanoplates, they could not be firmly adsorbed on the surface of the friction pair. Due to the layered structure of the nanoplate, some thin nanosheets would be peeled off from the nanoplates during the continuous extrusion process by the friction pair. These stripped nanosheets would be adsorbed on the surface of the friction pair and then form a lubricating film. However, due to the different thickness of the stripped nanosheets, the resulting lubricating film was uneven. In contrast, after ultrasonic dispersion, WS_2 nanoflowers were decomposed into some ultrathin nanosheets, as demonstrated by the TEM image in Figure 2f. We have researched the antifriction and antiwear mechanism of the ultrathin WS_2 nanosheets as additives in 500 SN base oil [38]. The antifriction mechanism of the WS_2 nanoflowers and the WS_2 ultrathin nanosheets is the same. When the nanoflowers were dispersed into nanosheets, the dispersed nanosheets quickly adhered to the surface of the friction pair and formed a lubricating film, further reducing the wear on the surface of the friction pair. Since the thickness of the ultrathin nanosheets forming the nanoflowers is substantially the same, when the nanoflowers are decomposed, a tribo-film with uniform thickness is formed. The uniform tribo-film can improve tribological performance. Therefore, WS_2 nanoflowers as a lubricant additive have better antifriction and antiwear properties than WS_2 nanoplates.

4. Conclusions

In this study, WS_2 hexagonal nanoplates and nanoflowers were successfully synthesized by a solid-phase reaction. Tribological tests demonstrated that the tribological properties of paraffin oil could be greatly improved with the addition of the two kinds of WS_2 nanomaterials, and the morphology and content of the WS_2 nanomaterials had a significant effect on the tribological properties of paraffin oil. The optimum nanomaterial concentration was 0.5 wt %. The paraffin oil with added WS_2 nanoflowers exhibited better friction reducing and antiwear properties than the WS_2 hexagonal nanoplates. With the addition of the WS_2 nanoflowers, the friction coefficient was stably maintained at a low value and the wear surface appeared to be smoother. The superior tribological performance of WS_2 nanoflowers can be attributed to their special structure. Since the nanoflowers are decomposed into a number of ultrathin nanosheets, and these nanosheets are adsorbed on the surface of the friction pair which forms a uniform tribo-film, this can reduce friction and wear.

Author Contributions: X.Z. and X.Y. designed the experiments. X.Z., J.W., and H.X. performed the experiments. X.Z. and H.T. analyzed the data. X.Z. and X.Y. wrote the manuscript. All authors read and approved the final manuscript.

Funding: This research was supported by the Jiangsu Province Industry-University-Research Cooperation Project (BY2018314), the Scientific Research Foundation of Jiangsu University of Technology (KYY18030) and Jiangsu Overseas Visiting Scholar Program for University Prominent Young & Middle-aged Teachers and Presidents.

Conflicts of Interest: The authors declare no conflict of interest.

References

1. Holmberg, K.; Andersson, P.; Nylund, N.O.; Makela, K.; Erdemir, A. Global energy consumption due to friction in trucks and buses. *Tribol. Int.* **2014**, *78*, 94–114. [CrossRef]
2. Holmberg, K.; Andersson, P.; Erdemir, A. Global energy consumption due to friction in passenger cars. *Tribol. Int.* **2012**, *47*, 221–234. [CrossRef]
3. Holmberg, K.; Siilasto, R.; Laitinen, T.; Andersson, P.; Sberg, A. Global energy consumption due to friction in paper machines. *Tribol. Int.* **2013**, *62*, 58–77. [CrossRef]
4. Zhang, B.S.; Xu, B.S.; Xu, Y.; Gao, F.; Shi, P.J.; Wu, Y.X. Cu nanoparticles effect on the tribological properties of hydrosilicate powders as lubricant additive for steel–steel contacts. *Tribol. Int.* **2011**, *44*, 878–886. [CrossRef]
5. Padgurskas, J.; Rukuiza, R.; Prosyčevas, I.; Kreivaitis, R. Tribological properties of lubricant additives of Fe, Cu and Co nanoparticles. *Tribol. Int.* **2013**, *60*, 224–232. [CrossRef]
6. Choi, Y.; Lee, C.; Hwang, Y.; Park, M.; Lee, J.; Choi, C.; Jung, M. Tribological behavior of copper nanoparticles as additives in oil. *Curr. Appl. Phys.* **2009**, *9*, e124–e127. [CrossRef]
7. Jeyaprakash, N.; Sivasankaran, S.; Prabu, G.; Yang, C.H.; Alaboodi, A. Enhancing the tribological properties of nodular cast iron using multi wall carbon nano-tubes (MWCNTs) as lubricant additives. *Mater. Res. Express* **2019**, *6*, 045038. [CrossRef]
8. Berman, D.; Erdemir, A.; Sumant, A.V. Graphene: A new emerging lubricant. *Mater. Today* **2014**, *17*, 31–42. [CrossRef]
9. Min, C.; Zhang, Q.; Shen, C.; Liu, D.; Shen, X.; Song, H.; Zhang, K. Graphene oxide/carboxyl-functionalized multi-walled carbon nanotube hybrids: Powerful additives for water-based lubrication. *RSC Adv.* **2017**, *7*, 32574–32580. [CrossRef]
10. Song, H.; Wang, Z.; Yang, J. Tribological properties of graphene oxide and carbon spheres as lubricating additives. *Appl. Phys. A* **2016**, *122*, 933. [CrossRef]
11. Rapoport, L.; Fleischer, N.; Tenne, R. Fullerene-like WS2 nanoparticles: Superior lubricants for harsh conditions. *Adv. Mater.* **2003**, *15*, 651–655. [CrossRef]
12. Zhang, L.L.; Tu, J.P.; Wu, H.M.; Yang, Y.Z. WS2 nanorods prepared by self-transformation process and their tribological properties as additive in base oil. *Mater. Sci. Eng. A* **2007**, *454*, 487–491. [CrossRef]
13. Zhang, X.; Xue, Y.; Ye, X.; Xu, H.; Xue, M. Preparation, characterization and tribological properties of ultrathin MoS2 nanosheets. *Mater. Res. Express* **2017**, *4*, 115011. [CrossRef]
14. Zhang, X.; Xue, M.; Yang, X.; Luo, G.; Yang, F. Hydrothermal synthesis and tribological properties of MoSe2 nanoflowers. *Micro Nano Lett.* **2015**, *10*, 339–342. [CrossRef]
15. Yang, J.H.; Yao, H.X.; Liu, Y.Q.; Wei, M.B.; Liu, Y.; Zhang, Y.J.; Wang, Y.X. Tribological properties of WSe2 nanorods as additives. *Cryst. Res. Technol.* **2009**, *44*, 967–970. [CrossRef]
16. Battez, A.H.; González, R.; Viesca, J.L.; Fernández, J.E.; Fernández, J.D.; Machado, A.; Riba, J. CuO, ZrO$_2$ and ZnO nanoparticles as antiwear additive in oil lubricants. *Wear* **2008**, *265*, 422–428. [CrossRef]
17. Peng, D.X.; Chen, C.H.; Kang, Y.; Chang, Y.P.; Chang, S.Y. Size effects of SiO$_2$ nanoparticles as oil additives on tribology of lubricant. *Ind. Lubr. Tribol.* **2010**, *62*, 111–120. [CrossRef]
18. Zhou, J.; Wu, Z.; Zhang, Z.; Liu, W.; Dang, H. Study on an antiwear and extreme pressure additive of surface coated LaF3 nanoparticles in liquid paraffin. *Wear* **2001**, *249*, 333–337. [CrossRef]
19. Zhao, C.; Chen, Y.K.; Jiao, Y.; Loya, A.; Ren, G.G. The preparation and tribological properties of surface modified zinc borate ultrafine powder as a lubricant additive in liquid paraffin. *Tribol. Int.* **2014**, *70*, 155–164. [CrossRef]
20. Gu, K.; Chen, B.; Chen, Y. Preparation and tribological properties of lanthanum-doped TiO$_2$ nanoparticles in rapeseed oil. *J. Rare Earth.* **2013**, *31*, 589–594. [CrossRef]
21. Braga, D.; Gutiérrez Lezama, I.; Berger, H.; Morpurgo, A.F. Quantitative determination of the band gap of WS2 with ambipolar ionic liquid-gated transistors. *Nano Lett.* **2012**, *12*, 5218–5223. [CrossRef] [PubMed]
22. Georgiou, T.; Jalil, R.; Belle, B.D.; Britnell, L.; Gorbachev, R.V.; Morozov, S.V.; Kim, Y.; Gholinia, A.; Haigh, S.J.; Makarovsky, O.; et al. Vertical field-effect transistor based on graphene–WS2 heterostructures for flexible and transparent electronics. *Nat. Nanotechnol.* **2013**, *8*, 100–103. [CrossRef]
23. Ansari, M.Z.; Ansari, S.A.; Parveen, N.; Cho, M.H.; Song, T. Lithium ion storage ability, supercapacitor electrode performance, and photocatalytic performance of tungsten disulfide nanosheets. *New J. Chem.* **2018**, *42*, 5859–5867. [CrossRef]

24. Liu, Z.; Li, N.; Su, C.; Zhao, H.; Xu, L.; Yin, Z.; Li, J.; Du, Y. Colloidal synthesis of 1T'phase dominated WS2 towards endurable electrocatalysis. *Nano Energy* **2018**, *50*, 176–181. [CrossRef]

25. Roy, S.; Bermel, P. Electronic and optical properties of ultra-thin 2D tungsten disulfide for photovoltaic applications. *Sol. Energ. Mat. Sol. C* **2018**, *174*, 370–379. [CrossRef]

26. Huang, S.; Wang, Y.; Hu, J.; Lim, Y.V.; Kong, D.; Zheng, Y.; Ding, M.; Pam, M.E.; Yang, H.Y. Mechanism Investigation of High-Performance Li–Polysulfide Batteries Enabled by Tungsten Disulfide Nanopetals. *ACS Nano* **2018**, *12*, 9504–9512. [CrossRef] [PubMed]

27. Ren, J.; Wang, Z.; Yang, F.; Ren, R.P.; Lv, Y.K. Freestanding 3D single-wall carbon nanotubes/WS2 nanosheets foams as ultra-long-life anodes for rechargeable lithium ion batteries. *Electrochim. Acta* **2018**, *267*, 133–140. [CrossRef]

28. Shang, X.; Chi, J.Q.; Lu, S.S.; Dong, B.; Li, X.; Liu, Y.R.; Yan, K.L.; Gao, W.K.; Chai, Y.M.; Liu, C.G. Novel CoxSy/WS2 nanosheets supported on carbon cloth as efficient electrocatalyst for hydrogen evolution reaction. *Int. J. Hydrog. Energy* **2017**, *42*, 4165–4173. [CrossRef]

29. Hu, K.H.; Wang, J.; Schraube, S.; Xu, Y.F.; Hu, X.G.; Stengler, R. Tribological properties of MoS2 nano-balls as filler in polyoxymethylene-based composite layer of three-layer self-lubrication bearing materials. *Wear* **2009**, *266*, 1198–1207. [CrossRef]

30. Wu, J.; Zhai, W.S.; Jie, G.F. Preparation and tribological properties of tungsten disulfide hollow spheres assisted by methyltrioctylammonium chloride. *Tribol. Int.* **2010**, *43*, 1650–1658.

31. Lu, Z.; Cao, Z.; Hu, E.; Hu, K.; Hu, X. Preparation and tribological properties of WS2 and WS2/TiO2 nanoparticles. *Tribol. Int.* **2019**, *130*, 308–316. [CrossRef]

32. Xu, Z.Y.; Hu, K.H.; Han, C.L.; Hu, X.G.; Xu, Y.F. Morphological influence of molybdenum disulfide on the tribological properties of rapeseed oil. *Tribol. Lett.* **2013**, *49*, 513–524. [CrossRef]

33. Rabaso, P.; Ville, F.; Dassenoy, F.; Diaby, M.; Afanasiev, P.; Cavoret, J.; Vacher, B.; Le Mogne, T. Boundary lubrication: Influence of the size and structure of inorganic fullerene-like MoS2 nanoparticles on friction and wear reduction. *Wear* **2014**, *320*, 161–178. [CrossRef]

34. Zhang, X.; Lei, W.; Ye, X.; Wang, C.; Lin, B.; Tang, H.; Li, C. A facile synthesis and characterization of graphene-like WS$_2$ nanosheets. *Mater. Lett.* **2015**, *159*, 399–402. [CrossRef]

35. Vattikuti, S.P.; Byon, C.; Chitturi, V. Selective hydrothermally synthesis of hexagonal WS$_2$ platelets and their photocatalytic performance under visible light irradiation. *Superlattice Microst.* **2016**, *94*, 39–50. [CrossRef]

36. Pang, Q.; Gao, Y.; Zhao, Y.; Ju, Y.; Qiu, H.; Wei, Y.; Chen, G. Improved Lithium-Ion and Sodium-Ion Storage Properties from Few-Layered WS$_2$ Nanosheets Embedded in a Mesoporous CMK-3 Matrix. *Chem. Eur. J.* **2017**, *23*, 7074–7080. [CrossRef] [PubMed]

37. Wu, Z.; Fang, B.; Bonakdarpour, A.; Sun, A.; Wilkinson, D.P.; Wang, D. WS$_2$ nanosheets as a highly efficient electrocatalyst for hydrogen evolution reaction. *Appl. Catal. B-Environ.* **2012**, *125*, 59–66. [CrossRef]

38. Zhang, X.; Xu, H.; Wang, J.; Ye, X.; Lei, W.; Xue, M.; Li, C. Synthesis of Ultrathin WS$_2$ Nanosheets and Their Tribological Properties as Lubricant Additives. *Nanoscale Res. Let.* **2016**, *11*, 442. [CrossRef] [PubMed]

Article

Interlayer Difference of Bilayer-Stacked MoS$_2$ Structure: Probing by Photoluminescence and Raman Spectroscopy

Xiangzhe Zhang [1], Renyan Zhang [1,*], Xiaoming Zheng [2], Yi Zhang [2], Xueao Zhang [3,*], Chuyun Deng [2,*], Shiqiao Qin [1,*] and Hang Yang [2]

[1] College of Advanced Interdisciplinary Research, National University of Defense Technology, Changsha 410073, China; xiangzhe.cheung@gmail.com

[2] College of Arts and Science, National University of Defense Technology, Changsha 410073, China; 162201014@csu.edu.cn (X.Z.); zhangyi1983@zju.edu.cn (Y.Z.); yanghangnudt@163.com (H.Y.)

[3] College of Physical Science and Technology, Xiamen University, Xiamen 361000, China

* Correspondence: ryancms@sina.cn (R.Z.); xazhang@nudt.edu.cn (X.Z.); dengchuyun@nudt.edu.cn (C.D.); sqqin8@nudt.edu.cn (S.Q.)

Received: 16 April 2019; Accepted: 17 May 2019; Published: 24 May 2019

Abstract: This work reports the interlayer difference of exciton and phonon performance between the top and bottom layer of a bilayer-stacked two-dimensional materials structure (BSS). Through photoluminescence (PL) and Raman spectroscopy, we find that, compared to that of the bottom layer, the top layer of BSS demonstrates PL redshift, Raman E_{2g}^1 mode redshift, and lower PL intensity. Spatial inhomogeneity of PL and Raman are also observed in the BSS. Based on theoretical analysis, these exotic effects can be attributed to substrate-coupling-induced strain and doping. Our findings provide pertinent insight into film–substrate interaction, and are of great significance to researches on bilayer-stacked structures including twisted bilayer structure, Van der Waals hetero- and homo-structure.

Keywords: film–substrate interaction; photoluminescence; Raman spectroscopy; molybdenum disulfide; bilayer-stacked structure

---

1. Introduction

By stacking up two single-layer two-dimensional (2D) materials, bilayer Van der Waals (VdW) homo- and hetero-structures can be fabricated [1]. Owing to the existence of interlayer coupling, these bilayer-stacked structures usually exhibit distinct properties from their monolayer counterparts. For example, energy band gap evolution is found in bilayer VdW homo-structures compared to the corresponding monolayer, as previously reported in graphene and MoS$_2$ [2,3]. Additionally, interlayer-coupling-induced p–n junction in VdW hetero-structure can lead to novel optoelectric effects [4,5]. Further, if stacking up two films with a misorientation angle, a brand-new tunable dimension is introduced to the bilayer-stacked two-dimensional materials structure (BSS), such BSS is referred to as twisted bilayer structure (tBLS). As a result, numerous exotic effects, induced by the twisted dimension and distinct from those in monolayer or bilayer without twisted angle, are expected. tBLS are tunable in their properties with variation in the twisted angle, thus have attracted intensive researches. For one thing, phonon in tBLS can be affected by interlayer coupling varying with angle, providing a simple but effective way to tune diverse properties such as, crystalline asymmetry [6], nonlinear optical effects [6,7], Raman scattering [8,9], and thermal conductivity [10–15]. For another, periodical interlayer Van der Waals potential can impact carrier performance of tBLS, which is first confirmed by the observation of Moiré pattern of twisted bilayer graphene (tBLG) under scanning tunneling microscope in 2005 [16]. Van Hove Singularity (VHS) [17] and angle-dependent electrical conductivity [18] are another two examples for this effect. Moreover, cutting-edge advances on twisted bilayer structure (tBLS) like, unconventional superconductivity in magic-angle tBLG [19] and mirror Dirac cone in incommensurate-angle tBLG [20], imply that there remains a lot that is yet to be explored.

However, all these findings about BSS focus only on the interlayer-coupling-induced effects, while ignoring the difference between the top and bottom layer of BSS. The BSS sample fabricated by transfer method can be divided into three different regions: stacked region where top and bottom layer overlap each other, bottom region (bottom layer excluding stacked region), and top region (top layer excluding stacked region). Although bottom and top regions are both supposed to be in direct contact with the substrate, there exists great difference between the top-substrate and the bottom-substrate coupling. Substrate coupling can affect 2D materials in many aspects. For example, on the one hand, substrate contact can employ strain on 2D materials, leading to phonon variations measured by Raman spectroscopy [21–23]. On the other hand, substrate can provide or deplete carriers depending on its doping type [24–26], thus tuning electrical and optical properties of materials deposited on it [27]. Furthermore, substrates with different permittivities and surface polar phonon modes demonstrate different scattering mechanisms limiting the electron mean free paths and mobility in 2D materials [28–30]. Band gap of semiconductors can also be tuned by dielectric environment permittivities [31]. Consequently, bottom and top region may demonstrate different phonon and exciton performance.

As one sort of transitional metal dichalcogenide (TMD) materials, MoS_2 monolayer with a two-dimensional structure demonstrates intriguing effects in various aspects, including optical [32–34], electrical [35–37], and thermal properties [38–40]. Especially, due to its unique direct band gap [41], monolayer MoS_2 is expected to have strong photoluminescence (PL) emission, which has been confirmed both experimentally and theoretically. Excited by 532 nm laser at ambient conditions, monolayer MoS_2 is reported to have two prominent PL peaks at 625 nm (B peak) and 670 nm (A peak) [42–44]. These two peaks correspond to two direct excitonic transitions at the Brillouin zone K point, while the difference between them comes from the spin-orbital coupling caused by valence band energy splitting [42]. Also, two easily identified Raman peaks are observed in MoS_2 monolayer [32,45], located near 390 cm^{-1} (E_{2g}^1, in-plane vibration mode) and 409 cm^{-1} (A_{1g}, out-of-plane vibration mode), respectively.

Through photoluminescence and Raman spectroscopy, we found that, in bilayer-stacked MoS_2 (BSM) samples fabricated by transfer, exciton and phonon performance in the top and bottom regions are remarkably different. Despite the fact that both top and bottom layers of BSM are transferred, compared to the bottom region, the top region demonstrates PL intensity reduction and peak redshift, implying less p-doping to top region from substrate. Meanwhile, redshift of in-plane Raman mode E_{2g}^1 is observed in top region, suggesting that vibration softens in top region. To exclude the film–substrate interaction, freestanding monolayer MoS_2 samples are fabricated. It is found that, compared to the supported region, the suspended region of monolayer MoS_2 demonstrates redshift in PL and Raman peaks, which are consistent with those in the top region of BSM, thus providing evidence for coupling difference between top-substrate and bottom-substrate.

Since the interlayer difference in BSS can complicate the experimental results, and is also affected by interlayer coupling, our findings are of great significance to distinguish between contributions from interlayer coupling and film–substrate interaction, which is of great significance to researches on interlayer-coupling-induced effects like optoelectric effects in VdW hetero-structure and twisted angle dependence in tBLS. Furthermore, our findings are universal and, apart from MoS_2 bilayer-stacked structure on SiO_2 substrate, we are sure this work can be informative to film–substrate interaction study on other 2D materials and substrates.

2. Materials and Methods

2.1. Sample Preparation

For this study, a convenient fabrication process is employed to obtain BSM samples. First, all single-layer flakes are deposited on a SiO_2/(001)Si substrate (SiO_2 layer is 300 nm thick), with a size of approximately 50 μm by chemical vapor deposition (CVD). Then, we transferred two sheets of monolayer MoS_2 to one substrate, by which method we can obtain tens of BSM samples with various angles in a single step. It should be noted that, in this work, both top and bottom layer of the BSM undergo transfer process to avoid preparation method induced difference. During the transfer process, any solvent that may cause doping in MoS_2 was avoided. Universally used transfer methods like, PMMA-way (poly-methyl-methacrylate) [46] and PVA-way (poly-vinyl-alcohol) [47] introduce contamination or wrinkles to the surface of materials. Herein, a previously reported PLLA-way (poly-L-lactic-acid) [48] is chosen to ensure the transfer is residual-free and of high-uniformity. After transfer, the as-fabricated samples undergo ultraviolet treatment [49] and annealing [50] (in a tube furnace in Ar/H_2 flow at 300 °C for 2 h) to remove residues and enhance interlayer coupling.

As for free-standing monolayer MoS_2 samples, they are fabricated by PLLA transfer onto SiO_2/(001)Si substrate (SiO_2 layer is 300 nm thick) with 300 nm-depth holes. These holes are of a radius 5 μm each, fabricated by reactive ion etching (RIE) method in SF_6/CHF_3 mixed gas flow (30 sccm).

2.2. Sample Characterization and Measurement

In this work, all bright field optical micrographs are taken by Nikon LV150 microscope, using 50× objective lens (Nikon, Tokyo, Japan). Dark-field optical micrographs are taken by ZEISS Axio Scope A1 microscope, using a 50× objective lens (Zeiss, Oberkochen, Germany). Atomic force microscopy (AFM) images are taken by NT-MDT Prima AFM system, using semi-contact scanning mode (NT-MDT, MoscowRussia). PL and Raman spectroscopy are measured by WITec Alpha300R confocal Raman system, using a 50× objective lens (WITec, Ulm, Germany). A 532 nm laser is used as the excitation source. For Raman measurements, the laser power is 1 mW, while for PL measurements laser power is 0.5 mW, sufficiently low to avoid heating effects. Optical gratings used for Raman and PL measurements are 1800 L/mm and 600 L/mm, respectively, providing respective spectral resolution smaller than 1 cm^{-1} and 1 nm.

3. Results

3.1. PL and Raman Difference between Layers of BSS

One as-fabricated BSM sample is shown in Figure 1. From the bright-field and dark-field optical micrographs in Figure 1a, we can see its surface is free of large-sized residual spots and wrinkles (of several micrometers size). To investigate its surface-height fluctuation in details, atomic force micrograph (AFM) is taken (Figure 1d). In Figure 1d, the sample's surface seems bubble-free, uniform, and plane within each region (no sharp morphology fluctuations of several micrometers size). At the edge between the stacked and top region, where top layer falls from bottom layer to substrate, there seems no ramp but a vertical cliff.

As is plotted in Figure 1b, bottom, stacked, and top region of this BSM sample demonstrate easily distinguishable PL intensity. Bottom region demonstrates the strongest PL intensity, then followed by the top region and stacked region in turn. Meanwhile, Figure 1c shows spatial inhomogeneity within the top region. The area in the vicinity of the stacked region (V-area), outlined by magenta dashed line, exhibits lower intensity than rest of the top region. Moreover, there appears a general correlation between PL and Raman over the mapped area, i.e., this V-area can also be easily identified in Raman intensity map and Raman shift map, as is shown in Figure 1e,f respectively. In this V-area, compared to the rest of the top region, Raman mode E^1_{2g} demonstrates redshift and intensity enhancement.

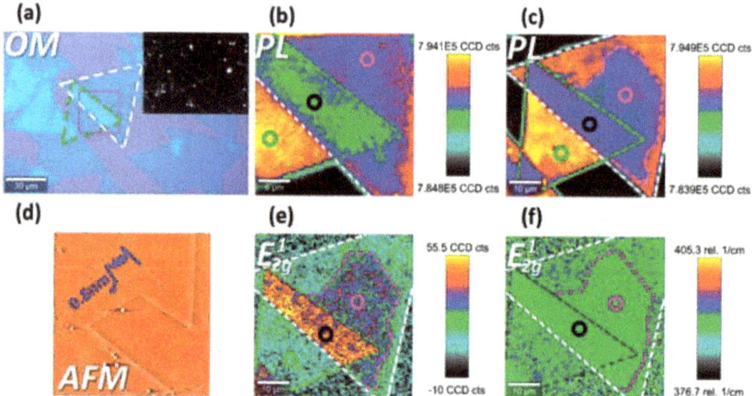

Figure 1. Interlayer difference of one twisted bilayer structure (tBLS) sample. Photoluminescence (PL) and Raman are all excited by 532 nm laser. (**a**) Optical micrograph. Inset corresponds to dark-field optical micrograph. The green and white dash lines outline the bottom and top layer respectively, while the red box outlines the scanning area in (**b**). (**b**,**c**) PL intensity map in high and low contrast respectively. (**d**) Atomic force microscopy (AFM) micrograph. (**e**) Raman intensity map of mode E^1_{2g}. (**f**) Raman shift map of mode E^1_{2g}. Green, black, and magenta circles in (**b**–**f**) point out bottom, stacked, and top region respectively.

This difference in PL and Raman spectra between the top and bottom regions is also observed in other as-fabricated BSM samples, as is shown in Figure 2. For all samples in Figure 2b, the maximum E^1_{2g}-to-A_{1g} Raman shift difference among the top region and bottom region is below 19 cm^{-1}, which is the signature of monolayer MoS$_2$, indicating these samples are stacked by two individual monolayer MoS$_2$. For each sample, compared to the bottom region, the top region exhibits PL intensity reduction and redshift (Figure 2a), and E^1_{2g} Raman mode redshift (Figure 2b), indicating this interlayer difference in all samples shares a common origin.

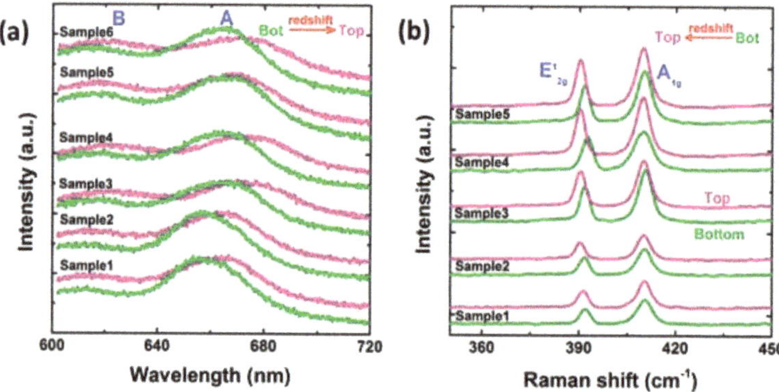

Figure 2. Photoluminescence and Raman spectra comparison between bottom (green) and top (magenta) layers. Bot and Top refer to bottom and top region, respectively. (**a**) PL spectra comparison. For clarity, spectra of one same tBLS sample are shifted vertically in small gap while spectra of different tBLS samples in large gap. Peak A and B are labeled. (**b**) Raman spectra comparison. Mode E^1_{2g} and A_{1g} are labeled.

3.2. Spatial Inhomogeneity in BSS

In addition, apart from the difference between the top and bottom layer, spatial inhomogeneity of PL emission and Raman scattering in the top region is prevalent among various BSM samples. Most importantly, it is found that, in many samples, area with lower PL intensity compared to the rest of the top region tends to emerge in the V-area. Another tBLS sample is shown in Figure 3, its PL intensity distribution on each region is consistent with the sample in Figure 1. It is noteworthy that an area (P2) outlined by pink dashed line in Figure 3c,e demonstrates identical PL intensity and E_{2g}^1 Raman shift with the V-area (P1) in this BSM sample. For more details, PL and Raman spectra on various regions are presented in Figure 3d,f respectively. Compared to the rest of the area of top region, PL emission of P1 and P2 demonstrates lower intensity and redshift, while Raman mode E_{2g}^1 also demonstrates redshift. Obviously, P1 and P2 are nearly same in PL and Raman spectra, implying identical exciton and phonon performance in these two regions. Moreover, though PL and Raman spectra in the V-area are remarkably different from rest of the top region (in Figures 1b–f and 3c,e), their corresponding AFM micrographs (Figures 1d and 3b) are spatially homogeneous. In contrast, the inhomogeneous area, with lower PL intensity, of bottom region (Figure 3c) matches exactly with the wrinkle and crack shown in the corresponding dark-field micrograph (Figure 3a inset). This implies that, the spatial inhomogeneity of PL and Raman spectra in top region is not due to abrupt variations in film morphology, like wrinkle, crack, and bubble.

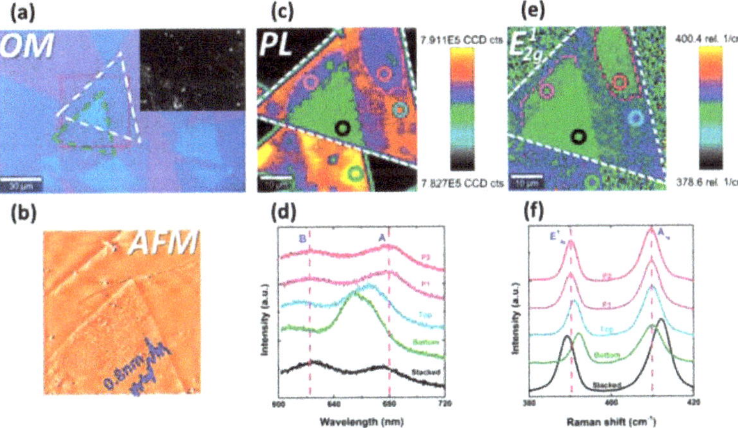

Figure 3. Photoluminescence and Raman inhomogeneity of tBLS. (**a**) Optical micrograph of one tBLS sample. Corresponding dark-field optical micrograph is shown in inset. Bottom and top layers are outlined by green and white dashed lines, respectively. Red box defines the scanning area of (**c,e**). (**b**) AFM micrograph of the same sample in (**a**). (**c**) PL intensity map. (**e**) Raman shift map of mode E_{2g}^1. (**d,f**) PL and Raman spectra of different regions. These regions are labeled by circles in corresponding colors in (**c,e**).

3.3. PL and Raman of Freestanding MoS$_2$

Interlayer difference and spatial inhomogeneity of PL and Raman spectroscopy in BSS possibly come from substrate-coupling difference between the layers, and among the top layer, respectively. To confirm this, we fabricated a freestanding sample (in Figure 4a) by transferring monolayer MoS$_2$ to SiO$_2$/Si substrate with holes of diameter 300 nm. The suspended (SUS) area of this sample totally excluded the film–substrate interaction. In Figure 4b–f, we can see that the suspended region demonstrates great intensity enhancement in PL and Raman spectroscopy, compared to that of the supported (SUP) region. This can be attributed to constructive interference effect in the top region [51]. Moreover, compared to the supported region, the suspended region demonstrates redshift in PL

(Figure 4c,d) and Raman peaks (Figure 4g–i). This implies less p-doping and vibration mode softens in suspended region, which is consistent with the top region, P1 (V-area), and P2 in BSM. Therefore, similar to suspended region, we can assume that top region, V-area, and P2 might be less affected by substrate contact.

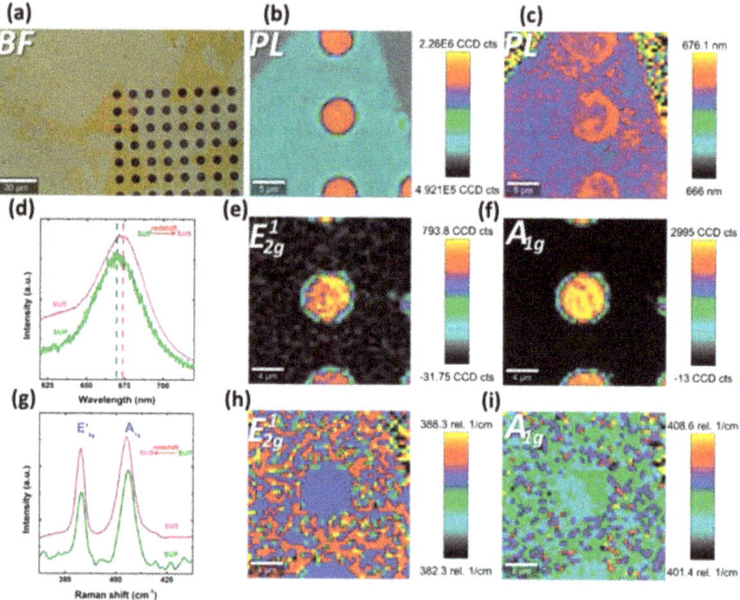

Figure 4. Free-standing monolayer MoS_2. (**a**) Bright-field optical micrograph. Red box outlines the scanning area of the middle and right column. (**b**) PL intensity map. (**c**) PL shift map. PL and Raman spectra in (**d**,**g**) are normalized for clarity. (**d**) PL spectra of suspended and supported region. SUP and SUS refer to supported and suspended region, respectively. (**e**,**f**) Raman intensity map of E_{2g}^1 and A_{1g}, respectively. (**g**) Raman spectra of suspended and supported region. (**h**,**i**) Raman shift map of E_{2g}^1 and A_{1g}, respectively.

4. Discussion

A schematic illustration is shown in Figure 5a. According to our findings in Figure 4, less p-doping and vibration mode softening are observed in suspended region compared to its substrate-supported counterpart, which are also observed in top region, V-area, and P2 in BSM. Therefore, we speculate that, while bottom region and supported region are in strong coupling with the substrate, top region just like the suspended region is in intermediate or weak coupling with the substrate, thus leading to less carrier transfer and strain from substrate. In transferred-fabricated samples, this coupling mainly comes from Van der Waals bonding instead of chemical bonding [26,52]. In addition, as shown in Figure 5, there might be film morphology fluctuations in the top region, including ripple formed by strain and stair at the edge of the bottom region. Stair and ripple correspond to P1 (V-area) and P2 region in Figure 3 respectively. Though these film morphology fluctuations might be less than one nanometer (the order of monolayer MoS_2 thickness), not sufficiently macroscopic to be detected by bright-field/dark-field optical microscope and atomic force microscopy, they can remarkably reduce film–substrate coupling.

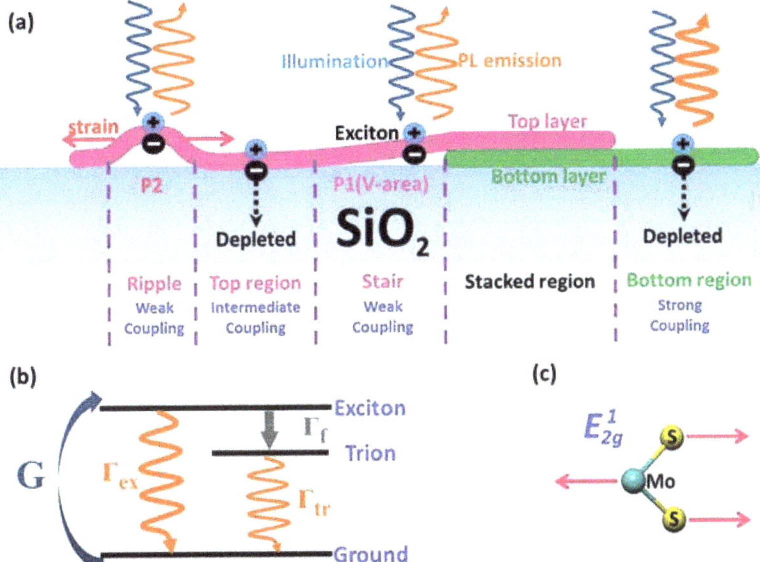

Figure 5. Schematic illustration. (**a**) Film–substrate coupling difference among bilayer-stacked two-dimensional materials structure (BSS) films. (**b**) Three-level energy diagram including exciton, trion, and ground. G represents the generation rate of exciton. Γ_{ex}, Γ_f, and Γ_{tr} represent exciton decay rate without trion formation rate, trion formation rate, and trion decay rate, respectively. (**c**) Schematic of in-plane Raman mode E_{2g}^1 for monolayer MoS_2.

For one thing, PL emission of MoS_2 is resulted from exciton (radiative wavelength: ~660 nm) and trion recombination (radiative wavelength: ~680 nm) [53]. According to related studies [24–26], the contribution ratio of exciton against trion determines the intensity and position of peak A. Since monolayer MoS_2 is an n-type semiconductor, the silicon oxide depletes equilibrium electrons in regions of strong coupling with substrate [25], which stabilize the radiative recombination process of exciton (Γ_{ex}) while suppressing the trion formation rate (Γ_f) at the same time, as is shown in Figure 5b. PL emission, in regions of strong coupling with substrate (e.g., bottom region), is exciton-dominant, thus demonstrating intensity enhancement and blueshift. In contrast, areas of less coupling with substrate, such as top region, ripple, and stair, where exciton contribution is reduced, are supposed to demonstrate lower intensity and redshift of PL.

For another, the in-plane Raman mode E_{2g}^1 corresponds to Mo and S atoms oscillating in the anti-phase parallel to the crystal plane, as shown in Figure 5c. As previously reported, E_{2g}^1 mode demonstrates redshift with uniform tensile uniaxial strain [21,22]. At the same time, it has been reported that film morphology fluctuations that lead to less coupling with substrate, like wrinkle and bubble, yield uniaxial tensile strain [54,55]. As a result, compared to the bottom region of strong substrate coupling, the top region of less substrate coupling is supposed to demonstrate E_{2g}^1 mode redshift caused by tensile uniaxial strain. Especially in ripple and stair regions, E_{2g}^1 is expected to demonstrate the strongest redshift.

The discussions above provide a reliable explanation for our findings. Admittedly, interference effects can induce intensity change of PL and Raman. However, on the one hand, the minor height fluctuations in BSS film cannot result in remarkable interference variations. On the other hand, interference-induced intensity change would be broad band, which is not consistent with our experimental results. Therefore, we conclude that substrate-coupling-induced strain and doping to BSS play the dominant part in interlayer difference and spatial inhomogeneity of phonon and exciton performance.

5. Conclusions

In summary, we conducted a systematic investigation on PL and Raman spectroscopy of bilayer-stacked MoS_2 fabricated by the transfer method. PL and Raman spectroscopy of freestanding monolayer MoS_2 are also measured for comparison. Interlayer difference and spatial inhomogeneity of exciton and phonon performance are experimentally observed in as-fabricated BSS samples, which we attribute to film–substrate coupling-induced strain and doping. Additionally, our findings prove that, even surface fluctuations less than one-atom-layer thickness can be easily identified by Raman and PL spectroscopy. This work will be of great use to inform future researches on BSS including tBLS, VdW homostructure and heterostructure, and improve our understanding of substrate effects on optical and transport properties of 2D materials.

Author Contributions: X.Z. (Xiangzhe Zhang) and R.Z. conceived and designed the experiments; X.Z. (Xiangzhe Zhang), X.Z. (Xiaoming Zheng), Y.Z. performed the experiments; S.Q., C.D., R.Z., H.Y. and X.Z. (Xueao Zhang) provided valuable suggestions; X.Z. (Xiangzhe Zhang) wrote the paper.

Funding: This work was supported by National Natural Science Foundation (NSF) of China (Grant No. 11802339, 11805276, 61805282, 61801498, 11804387, 11404399, 11874423, 51701237); Scientific Researches Foundation of National University of Defense Technology (Grant No. ZK16-03-59, ZK18-01-03, ZK18-03-36, ZK18-03-22); NSF of Hunan province (Grants No.2016JJ1021); Open Director Fund of State Key Laboratory of Pulsed Power Laser Technology (SKL2018ZR05); Open Research Fund of Hunan Provincial Key Laboratory of High Energy Technology (Grant No. GNJGJS03); Opening Foundation of State Key Laboratory of Laser Interaction with Matter (Grant No. SKLLIM1702); Youth talent lifting project (Grant No. 17-JCJQ-QT-004).

Acknowledgments: The authors would like to acknowledge professor Gang Peng for constructive suggestions. We also acknowledge Yuehua Wei for providing materials.

Conflicts of Interest: The authors declare no conflicts of interest.

References

1. Geim, A.K.; Grigorieva, I.V. Van der Waals heterostructures. *Nature* **2013**, *499*, 419–425. [CrossRef] [PubMed]
2. Lopezsanchez, O.; Lembke, D.; Kayci, M. Ultrasensitive photodetectors based on monolayer MoS_2. *Nat. Nanotechnol.* **2013**, *8*, 497. [CrossRef]
3. Zande, A.M.V.D.; Kunstmann, J.; Chernikov, A. Tailoring the electronic structure in bilayer molybdenum disulfide via interlayer twist. *Nano Lett.* **2014**, *14*, 3869–3875. [CrossRef] [PubMed]
4. Lee, C.H.; Lee, G.H.; Zande, A.M.V.D. Atomically thin p-n junctions with van der Waals heterointerfaces. *Nat. Nanotechnol.* **2014**, *9*, 676–681. [CrossRef] [PubMed]
5. Wang, F.; Wang, Z.; Xu, K. Tunable GaTe-MoS2 van der Waals p-n Junctions with Novel Optoelectronic Performance. *Nano Lett.* **2015**, *15*, 7558–7566. [CrossRef] [PubMed]
6. Shan, Y.; Li, Y.; Huang, D. Stacking-symmetry governed second harmonic generation in graphene trilayers. *Sci. Adv.* **2018**, *4*, eaat0074. [CrossRef]
7. Hsu, W.T.; Zhao, Z.A.; Li, L.J. Second harmonic generation from artificially stacked transition metal dichalcogenide twisted bilayers. *Acs Nano* **2014**, *8*, 2951–2958. [CrossRef]
8. Havener, R.W.; Zhuang, H.; Brown, L. Angle-resolved Raman imaging of interlayer rotations and interactions in twisted bilayer graphene. *Nano Lett.* **2012**, *12*, 3162–3167. [CrossRef]
9. Kim, K.; Coh, S.; Tan, L.Z. Raman Spectroscopy Study of Rotated Double-Layer Graphene: Misorientation-Angle Dependence of Electronic Structure. *Phys. Rev. Lett.* **2012**, *108*, 246103. [CrossRef]
10. Balandin, A.A.; Ghosh, S.; Bao, W. Superior thermal conductivity of single-layer graphene. *Nano Lett.* **2008**, *8*, 902. [CrossRef]
11. Nika, D.L.; Pokatilov, E.P.; Askerov, A.S. Phonon thermal conduction in graphene: Role of Umklapp and edge roughness scattering. *Phys. Rev. B* **2009**, *79*, 155413. [CrossRef]
12. Balandin, A.A. Thermal properties of graphene and nanostructured carbon materials. *Nat. Mater.* **2011**, *10*, 569. [CrossRef]

13. Cai, W.; Moore, A.; Chen, S. Thermal transport in suspended and supported monolayer graphene grown by chemical vapor deposition. *Nano Lett.* **2010**, *10*, 1645–1651. [CrossRef]

14. Li, H.; Ying, H.; Chen, X. Thermal conductivity of twisted bilayer graphene. *Nanoscale* **2014**, *6*, 13402–13408. [CrossRef] [PubMed]

15. Chenyang, L.; Bishwajit, D.; Xiaojian, T. Commensurate lattice constant dependent thermal conductivity of misoriented bilayer graphene. *Carbon* **2018**, *138*, 451–457.

16. Pong, W.T.; Durkan, C. TOPICAL REVIEW: A review and outlook for an anomaly of scanning tunnelling microscopy (STM): Superlattices on graphite. *J. Phys. D Appl. Phys.* **2005**, *38*, R329. [CrossRef]

17. Li, G.; Luican, A.; Santos, J.M.B.L.D. Observation of Van Hove singularities in twisted graphene layers. *Nat. Phys.* **2009**, *6*, 109–113. [CrossRef]

18. Liao, M.; Wu, Z.W.; Du, L. Twist angle-dependent conductivities across $MoS2$/graphene heterojunctions. *Nat. Commun.* **2018**, *9*, 4068. [CrossRef]

19. Cao, Y.; Fatemi, V.; Fang, S. Unconventional superconductivity in magic-angle graphene superlattices. *Nature* **2018**, *556*, 43–50. [CrossRef]

20. Yao, W.; Wang, E.; Bao, C. Quasicrystalline 30° twisted bilayer graphene as an incommensurate superlattice with strong interlayer coupling. *Talanta* **2018**, *184*, 50. [CrossRef]

21. Castellanos-Gomez, A.; Roldán, R.; Cappelluti, E. Local strain engineering in atomically thin MoS_2. *Nano Lett.* **2013**, *13*, 5361–5366. [CrossRef]

22. Su, L.; Yu, Y.; Cao, L. In Situ Monitoring of the Thermal-Annealing Effect in a Monolayer of MoS_2. *Phys. Rev. Appl.* **2017**, *7*, 034009. [CrossRef]

23. Su, L.; Yu, Y. Effects of substrate type and material-substrate bonding on high-temperature behavior of monolayer WS_2. *Nano Res.* **2015**, *8*, 2686–2697. [CrossRef]

24. Tongay, S.; Zhou, J.; Ataca, C. Broad-range modulation of light emission in two-dimensional semiconductors by molecular physisorption gating. *Nano Lett.* **2013**, *13*, 2831. [CrossRef]

25. Mouri, S.; Miyauchi, Y.; Matsuda, K. Tunable photoluminescence of monolayer MoS_2 via chemical doping. *Nano Lett.* **2013**, *13*, 5944–5948. [CrossRef]

26. Su, L.; Zhang, Y.; Yu, Y. Dependence of coupling of quasi 2-D $MoS2$ with substrates on substrate types, probed by temperature dependent Raman scattering. *Nanoscale* **2014**, *6*, 4920. [CrossRef]

27. Sercombe, D.; Schwarz, S.; Pozozamudio, O.D. Optical investigation of the natural electron doping in thin MoS_2 films deposited on dielectric substrates. *Sci. Rep.* **2013**, *3*, 3489. [CrossRef]

28. Giannazzo, F.; Sonde, S.; Nigro, R.L.; Rimini, E.; Raineri, V. Mapping the density of scattering centers limiting the electron mean free path in graphene. *Nano Lett.* **2011**, *11*, 4612–4618. [CrossRef]

29. Radisavljevic, B.; Radenovic, A.; Brivio, J. Single-layer MoS_2 transistors. *Nat. Nanotechnol.* **2011**, *6*, 147–150. [CrossRef]

30. Yu, Z.; Ong, Z.; Pan, Y. Realization of Room-Temperature Phonon-Limited Carrier Transport in Monolayer $MoS2$ by Dielectric and Carrier Screening. *Adv. Mater.* **2016**, *28*, 547–552. [CrossRef]

31. Giannazzo, F. Engineering 2D heterojunctions with dielectrics. *Nat. Electron.* **2019**, *2*, 54–55. [CrossRef]

32. Yue, N.; Sergio, G.A.; Riccardo, F. Thickness-Dependent Differential Reflectance Spectra of Monolayer and Few-Layer MoS_2, $MoSe_2$, WS_2 and WSe_2. *Nanomaterials* **2018**, *8*, 725.

33. Soh, D.B.S.; Rogers, C.; Gray, D.J. Optical nonlinearities of excitons in monolayer MoS_2. *Phys. Rev. B* **2017**, *97*, 165111. [CrossRef]

34. Ramasubramaniam, A. Large excitonic effects in monolayers of molybdenum and tungsten dichalcogenides. *Phys. Rev. B: Condens. Matter* **2012**, *86*, 2757–2764. [CrossRef]

35. Qiu, H.; Pan, L.; Yao, Z. Electrical characterization of back-gated bi-layer MoS_2 field-effect transistors and the effect of ambient on their performances. *Appl. Phys. Lett.* **2012**, *100*, 183.

36. Das, S.; Chen, H.Y.; Penumatcha, A.V. High performance multilayer $MoS2$ transistors with scandium contacts. *Nano Lett.* **2013**, *13*, 100–105. [CrossRef]

37. Mak, K.F.; Mcgill, K.L.; Park, J. The valley Hall effect in MoS_2 transistors. *Science* **2014**, *344*, 1489–1492. [CrossRef]

38. Sahoo, S.; Gaur, A.P.S.; Ahmadi, M. Temperature-Dependent Raman Studies and Thermal Conductivity of Few-Layer MoS_2. *Physics* **2013**, *117*, 9042–9047. [CrossRef]

39. Wu, L.; Carrete, J.; Mingo, N. Thermal conductivity and phonon linewidths of monolayer MoS_2 from first principles. *Appl. Phys. Lett.* **2013**, *103*, 109.

40. Liu, X.; Zhang, G.; Pei, Q.X. Phonon thermal conductivity of monolayer MoS$_2$ sheet and nanoribbons. *Appl. Phys. Lett.* **2013**, *103*, 1271. [CrossRef]
41. Hone, J. Atomically Thin MoS$_2$: A New Direct-Gap Semiconductor. *Phys. Rev. Lett.* **2010**, *105*, 136805.
42. Splendiani, A.; Sun, L.; Zhang, Y. Emerging photoluminescence in monolayer MoS$_2$. *Nano Lett.* **2010**, *10*, 1271–1275. [CrossRef] [PubMed]
43. Tonndorf, P.; Schmidt, R.; Bottger, P. Photoluminescence emission and Raman response of MoS$_2$, MoSe$_2$, and WSe$_2$ nanolayers. *Opt. Express* **2013**, *21*, 4908–4916. [CrossRef]
44. Mak, K.F.; He, K.; Lee, C. Tightly bound trions in monolayer MoS$_2$. *Nat. Mater.* **2013**, *12*, 207–211. [CrossRef] [PubMed]
45. Li, H.; Zhang, Q.; Yap, C.C.R. From Bulk to Monolayer MoS$_2$: Evolution of Raman Scattering. *Adv. Funct. Mater.* **2012**, *22*, 1385–1390. [CrossRef]
46. Van, N.H.; Qian, Y.; Han, S.K. PMMA-Etching-Free Transfer of Wafer-scale Chemical Vapor Deposition Two-dimensional Atomic Crystal by a Water Soluble Polyvinyl Alcohol Polymer Method. *Sci. Rep.* **2016**, *6*, 33096.
47. Marta, B.; Leordean, C.; Istvan, T. Efficient etching-free transfer of high quality, large-area CVD grown graphene onto polyvinyl alcohol films. *Appl. Surf. Sci.* **2016**, *363*, 613–618. [CrossRef]
48. Li, H.; Wu, J.; Huang, X. A universal, rapid method for clean transfer of nanostructures onto various substrates. *Acs Nano* **2014**, *8*, 6563. [CrossRef]
49. Yang, H.; Qin, S.Q.; Peng, G. Ultraviolet-Ozone treatment for effectively removing adhesive residue on graphene. *Nano* **2016**, *11*, 147. [CrossRef]
50. Woods, C.R.; Withers, F. Macroscopic self-reorientation of interacting two-dimensional crystals. *Nat. Commun.* **2016**, *7*, 10800. [CrossRef]
51. Gao, L.; Ren, W.; Liu, B. Surface and Interference Coenhanced Raman Scattering of Graphene. *ACS Nano* **2009**, *3*, 933–939. [CrossRef]
52. Wang, Y.Y.; Ni, Z.H.; Yu, T. Raman Studies of Monolayer Graphene: The Substrate Effect. *J. Phys. Chem. C* **2008**, *112*, 10637–10640. [CrossRef]
53. Ellis, J.K.; Lucero, M.J.; Scuseria, G.E. The indirect to direct band gap transition in multilayered MoS$_2$ as predicted by screened hybrid density functional theory. *Appl. Phys. Lett.* **2011**, *99*, 8207. [CrossRef]
54. Timoshenko, S.; Woinowsky-Krieger, S.; Woinowsky, S. *Theory of Plates and Shells*; McGraw-Hill: New York, NY, USA, 1959; Volume 2.
55. Landau, L.; Pitaevskii, L.; Lifshitz, E.; Kosevich, A. *Theory of Elasticity*, 3rd ed.; Butterworth-Heinemann: Oxford, UK, 1959; p. 195.

Article

Direct Observation of Raman Spectra in Black Phosphorus under Uniaxial Strain Conditions

Stacy Liang, Md Nazmul Hasan and Jung-Hun Seo *

Department of Materials Design and Innovation, University of Buffalo, Buffalo, NY 14260, USA;
stacyli@buffalo.edu (S.L.); mhasan9@buffalo.edu (M.N.H.)
* Correspondence: junghuns@buffalo.edu; Tel.: +1-716-645-1881

Received: 22 February 2019; Accepted: 25 March 2019; Published: 8 April 2019

Abstract: In this paper, we systematically studied the Raman vibration of black phosphorus (BP) transferred onto a germanium (Ge)-coated polydimethylsiloxane (PDMS) substrate, which generates a much higher contrast in BP. This engineered flexible substrate allowed us to directly observe a much thinner BP layer on the flexible substrate at the desired location. Therefore, it enabled us to perform Raman spectroscopy immediately after exfoliation. The Raman spectra obtained from several BP layers with different thicknesses revealed that the clear peak shifting rates for the A_g^{1}, B_{2g}, and A_g^{2} modes were 0.15, 0.11, and 0.11 cm^{-1}/nm, respectively. Using this value to identify a 2–3-layered BP, a study on the strain–Raman spectrum relationship was conducted, with a maximum uniaxial strain of 0.89%. The peak shifting of A_g^{1}, B_{2g}, and A_g^{2} caused by this uniaxial strain were measured to be 0.86, 0.63, and 0.21 $cm^{-1}/\Delta\varepsilon$, respectively.

Keywords: black phosphorus; uniaxial strain; flexible substrate

---

1. Introduction

Strain engineering has been known as an effective way to modulate the electronic, transport, and optical properties of semiconductors [1–4]. This method is particularly powerful when engineering low-dimensional semiconductors, such as one- and two-dimensional semiconductors (1D and 2D, respectively), since these low-dimensional semiconductors can tolerate much higher strain levels than three-dimensional (3D) semiconductors, such as bulk or thin-films [5,6]. For example, graphene is known to sustain strains of up to 15% without any noticeable damage to its crystalline structure [6,7]. As a result, strain engineering is a viable way to tune low-dimensional semiconductors' electrical, optical, chemical, and mechanical performances [8–12].

Recently, black phosphorus (BP, also known as phosphorene) was mechanically exfoliated from its bulk format [13–15]. Unlike the widely studied graphene, BP exhibits a finite and direct band gap varying from 0.5 eV in bulk to ~1.2 eV for a single layer, and its free-carrier mobility (approximately 1000 cm^2/v·s) is better than that of other typical 2D semiconductors, such as molybdenum sulfate (MoS_2; approximately 200 cm^2/v·s) [16–19]. As a result, various optoelectronic applications such as photodetectors and field-effect transistors, have been recently demonstrated [20–23]. Another attractive property of BP is its wide range of band gap tunability by mechanical strain. It is predicted that the band gap of BP can be modulated from as low as 0.55 eV up to 1.1 eV, while still maintaining a direct band gap by the application of ±8% of mechanical strain [24–26]. Such a wide tunability in the band gap of BP from 0.55 eV to 1.1 eV corresponds to the 1000 nm to 2200 nm wavelength range, suggesting that BP can be used as an active material for the near-infrared tunable light source. For this reason, there have been several theoretical and empirical attempts to demonstrate the characteristics of BP under strain conditions [27,28]. However, most of the experiments on the strain properties of BP have used thick BP (>10 nm thickness), because the visibility of BP decreases dramatically as it becomes thinner (similar to other 2D materials). For example, 13 nm- and 15 nm-thick BP (>20 layers) were used in the studies by Zhu et al. to examine the mechanical robustness of flexible BP devices under bending conditions. Material characterization studies of few-layered BP directly on a flexible substrate under strain conditions have not been performed so far.

In this paper, we investigated the strain dependence of the Raman characteristics of BP, taken directly from a flexible polymer substrate, enabled by a thin layer of germanium (Ge) coated on the backside of the polymer. This thin Ge contrast booster reflector on the backside of the polymer substrate provided a much higher contrast in BP and allowed us to investigate BP at the desired location with specific thicknesses. Also, our structure allowed us to perform Raman spectroscopy immediately after exfoliation and to perform Raman characterization faster avoiding the structural degradation caused by environmental factors, such as oxidation, which can potentially cause unintended Raman shifts.

2. Materials and Methods

Figure 1 shows a schematic illustration of the preparation of the BP sample. The BP sample (purity of 99.9999%) was purchased from 2D Semiconductors USA Inc (Scottsdale, AZ, USA). The process started with mechanical exfoliation from the bulk BP using a well-known micromechanical cleavage technique (also known as the "Scotch-tape" method), which is commonly used to create other 2D semiconductors from their bulk formats (Figure 1(i)) [29]. In this step, thin layers of BP that were a few nanometers thick were obtained. The thin BP layers were then carefully placed onto an ultrathin (>30 um) polydimethylsiloxane (PDMS) substrate prepared by spin-coating on a Petri dish (Figure 1(ii,iii)). Prior to the BP transfer step, a thin Ge layer (200nm) was deposited on the backside of the PDMS substrate using an e-beam evaporation method, which significantly enhanced the reflection of BP on the PDMS substrate. As shown in Figure 1(iv), once the transfer process was completed, the samples were immediately characterized under different strain conditions within 10 min to avoid unwanted BP degradation. Also, we prepared new samples each time we performed Raman spectroscopy under different strain conditions to maintain a high BP quality.

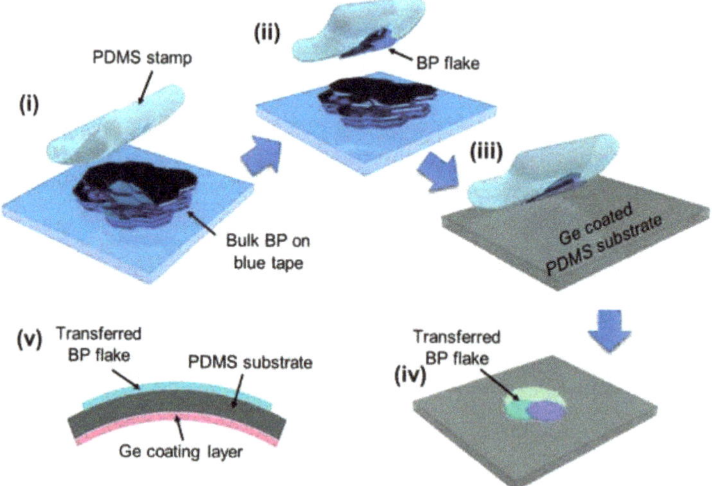

Figure 1. A schematic illustration of the sample preparation process on a Ge-coated polydimethylsiloxane (PDMS) substrate. (**i**) Mechanical exfoliation of black phosphorus (BP) flakes from bulk BP using blue tape, (**ii**) pick-up process of exfoliated BP flakes from the blue tape, (**iii,iv**) a transfer process of BP flakes onto the Ge-coated PDMS substrate, (**v**) cross-sectional view of the final structure under bending conditions.

3. Results and Discussion

3.1. Structure Analysis by Optical Simulation

In most of the 2D materials, it is well known that mono- or few-layered 2D materials are visible only when they are transferred onto a specific substrate that has particular refractive index and dielectric constant. Figure 2a,b show the relationship between the simulated reflectance versus wavelength of BP on a thin PDMS substrate (with and without Ge-coating on the PDMS substrate) as a function of BP thickness. To simulate the contrast enhancement of the BP layer obtained by employing the Ge-coated PDMS substrate and compare it with that of the reference bare PDMS substrate, we calculated the reflection of the multi-layered structure (BP/PDMS/Ge) using the Fresnel equation and Snell's equation [30]. The light reflected by an interface is determined by the discontinuity components of the two materials. For multiple interfaces, the total amount of reflected light is the sum of individual reflections. Depending upon their phase relationships, the reflections from the interfaces can be calculated using Equation (1) [30]:

$$R = \frac{(n-2)^2 + k^2}{(n+2)^2 + k^2} \tag{1}$$

where n is the refractive index, and k is the absorptance of the film. The optical contrast (C_λ) is the fractional change of reflection because of the presence of the Ge layer on the substrate and is defined by Equation (2) [31]:

$$C_\lambda = \frac{R_{BPonPDMS} - R_{Ge+BPonPDMS}}{R_{Ge+BPonPDMS}} \tag{2}$$

where $R_{Ge+BPonPDMS}$ and $R_{BPonPDMS}$ are the reflected intensities collected from the Ge-coated PDMS substrate and the reference PDMS substrate, respectively. Therefore, when we designed the substrate structure, we carefully checked the reflective index of each layer, since the visibility (i.e., the contrast of BP) can be enhanced or reduced depending on the reflective index (n). In our case, the refractive indices of 3.1 and 1.4 for BP and PDMS, respectively, were used in the calculation. However, any materials that have a high-reflective index, such as Si (n_{Si} = 4.1), can also enhance the contrast of BP, although Ge has better optical and process advantages (higher reflective index and easier to deposit at a lower temperature) compared to Si. The simulated reflectance clearly indicated that the reflectance of BP was nearly invisible (<5%) when the thickness of the BP was less than 3.5 nm (i.e., five layers), which agrees well with experimental observations [19]. On the other hand, as shown in Figure 2b, the simulated reflectance of BP on a thin Ge-coated PDMS substrate showed a significantly enhanced reflectance, namely, 37%, 20%, 9%, and 4.5% for 7, 3.5, 2.1, and 0.7 nm-thick BP, respectively. Interestingly, the peak wavelength shifted to a shorter wavelength as the thickness of BP was reduced. For example, the peak wavelength that appeared at 480 nm for the 7 nm-thick BP appeared at 410 nm for the 0.7 nm-thick BP. This simulated color shift was also observed in our experiment, as shown in Figure 2c. As shown in Figure 2d, an atomic force microscopy (AFM) analysis was carried out using a Bruker AFM system with non-contact mode from the 20 mm × 20 mm area to prevent any possible damage to the sample during surface profiling. The layer thickness profile also matched well with the simulated thickness–color relationship. In other words, the wavelength changed depending on the refractive index and thickness of the thin film: $2 \cdot t = m \cdot \lambda / n_{film}$, where λ is the wavelength of the reflected light, and m is an integer. When the refractive index is fixed, as in our case, this equation can be rewritten as $2 \cdot t = m \cdot \lambda / n_{film}$, which shows a proportional relationship between the thickness of a thin film (t) and the wavelength (λ). As simulated and experimentally verified, we observed the wavelength shift to a lower wavelength as the thickness of the BP layer decreased. This thickness–wavelength shifting relationship was stronger when the refractive index was higher.

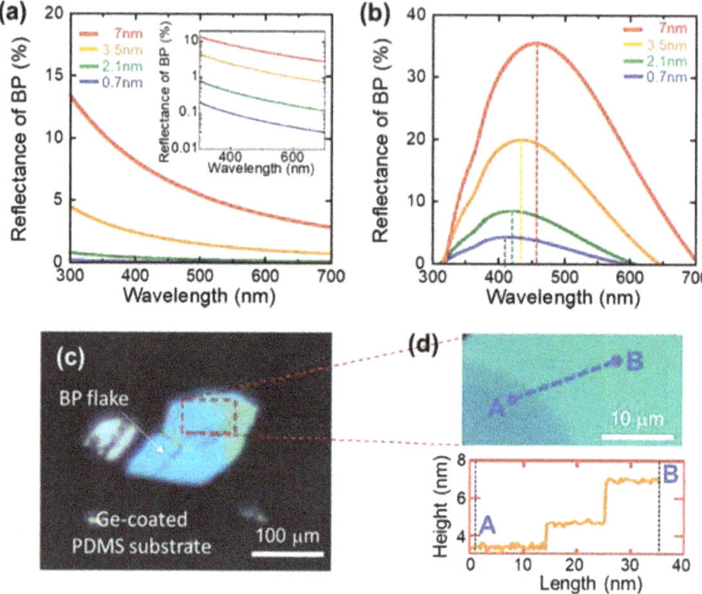

Figure 2. Simulated reflectance of BP on (**a**) a typical PDMS substrate (~30 um) and (**b**) a Ge-coated PDMS substrate (30 um + 200 nm). The inset of Figure 2a is the log-scale reflectance of BP to show the details of reflectance. (**c**) Microscopic image showing typical BP flakes on a Ge-coated PDMS substrate. (**d**) (lower) AFM surface profile between point A and B. (upper) The scanned area is shown.

Therefore, it is clear that the thin Ge-coated PDMS substrate improved the reflectance of BP enough to directly observe its crystal shapes under the microscope. Such high contrast also allowed us to find the same location in Raman spectroscopy during strain characterization, as described below.

3.2. Characterization of the Relationship between Raman Spectraand Thickness in BP

In order to investigate the crystalline quality of BP, Raman spectroscopy was performed using Renishaw Raman spectroscopy. The excitation was provided by a linearly polarized 514 nm excitation laser along a zigzag direction with a $50\times$ objective lens. The diameter of the laser spot was 1 μm. In order to avoid BP ablation caused by laser-induced heating, all Raman spectra were recorded at a low laser power (200 uW) with an exposure time of 10 s and accumulations of 10 times. Figure 3a shows the Raman spectra of transfer-printed BP ranging from 150 nm to 3 nm on a Ge-coated PDMS substrate. Figure 3a represents the Raman spectra of BP layers with different thicknesses ranging from 360 cm^{-1} to 480 cm^{-1}. In each spectrum, three Raman modes were present at 360 cm^{-1}, 436 cm^{-1}, and 464 cm^{-1}, each of which was assigned a unique phonon mode of BP: (1) A_g^1, (2) B_{2g}, (3) and A_g^2 [19,32,33]. Since the observed BP phonon modes matched the phonon modes seen in their bulk form (centered at 361 cm^{-1}, 438 cm^{-1}, and 466 cm^{-1}), the Raman spectra confirmed that the lattice of BP was retained during the exfoliation step. Also, when we performed Raman spectroscopy, we investigated the edge of the BP flake and confirmed the armchair direction by comparing the relative peak intensity of the A_g^1, B_{2g}, and A_g^2 phonon modes. Once we confirmed the direction of the BP edge, the sample was rotated 90 degrees and attached to the metal mold to perform a Raman spectroscopy under bending conditions. Figure 3b–d shows each phonon mode of BP as its thickness was reduced from 150 nm to 3 nm. All of the Raman phonon modes (A_g^1, B_{2g}, and A_g^2) demonstrated blue-shifting as the thickness increased. Figure 3b–d presents the overlay peak positions of the A_g^1, B_{2g}, and A_g^2 phonon modes as a function of the wavenumber, which showed a noticeable peak shifting. Figure 4a–c represents the trend in A_g^1, B_{2g}, and A_g^2 Raman peaks as a function of their thickness. The blue dotted lines in Figure 4 show the polynomial extrapolation of the measured data points. The peak wavenumbers of the A_g^1, B_{2g}, and A_g^2 modes reached 364 cm^{-1}, 442 cm^{-1}, and 469 cm^{-1}, as BP became a monolayer. The peak positions of the A_g^1, B_{2g}, and A_g^2 modes gradually decreased until the thickness of BP reached about 40–50 nm. As shown in Figure 4, all the A_g^1, B_{2g}, and A_g^2 phonon modes were saturated when thicker than 40–50 nm; therefore, the BP thickness of 40–50 nm was a transition point at which the bulk property became dominant. We also noticed that the three Raman modes (A_g^1, B_{2g}, and A_g^2) had different sensitivities to their thicknesses; in other words, the peak shifting rates for the A_g^1, B_{2g}, and A_g^2 modes were 0.15 cm^{-1}/nm, 0.11cm^{-1}/nm, and 0.11cm^{-1}/nm, respectively. The slightly higher shifting rate in the A_g^1 mode could be explained by the stiffer A_g^1 vibration with increasing BP thickness. Therefore, the peak distance between the A_g^1 modes and B_{2g} or A_g^2 modes, namely, the difference in their frequencies ($\Delta\omega$), could be used as an effective thickness indicator in order to examine the thickness of BP layers.

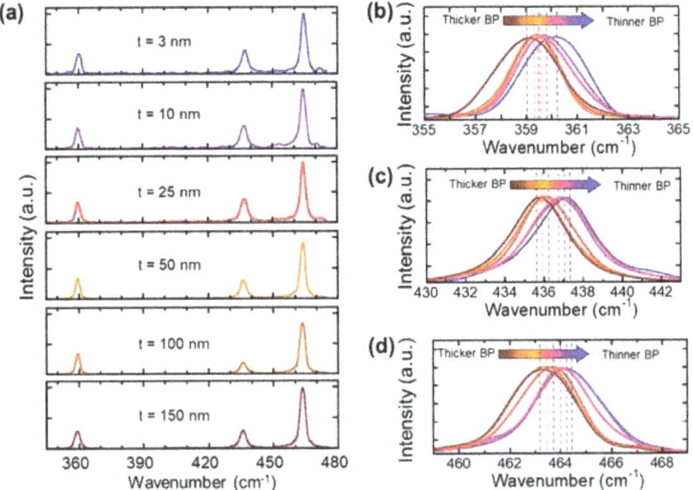

Figure 3. (a) Raman spectra of BP taken at different thicknesses from 150 nm to 3 nm. (**b–d**) Magnified Raman spectra of the A_g^1, B_{2g}, and A_g^2 Raman modes as a function of their thickness.

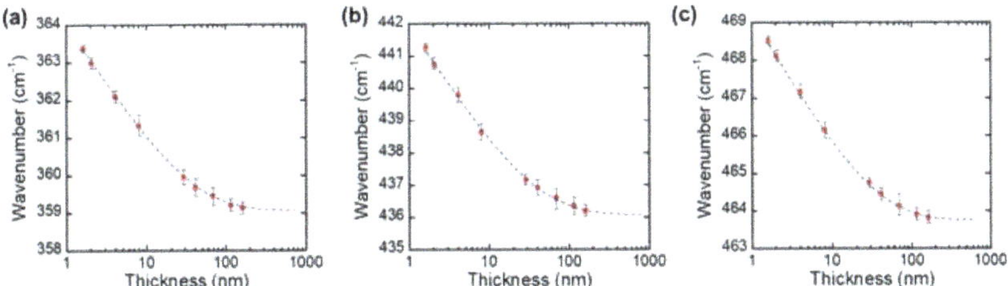

Figure 4. Peaks of (a) A_g^1, (b) B_{2g}, and (c) A_g^2 Raman modes as a function of their thickness.

3.3. Characterization of the Raman vs Strain Relationship of BP

After we confirmed the crystalline quality and thickness of BP layers, we performed a strain–Raman relationship spectral study to investigate the Raman shifts under different uniaxial strain conditions. On the basis of the results shown in Figures 2 and 3, the thickness of BP for the strain-Raman relationship spectrum study was found to correspond to 2–3 layers. In order to accurately measure the changes in the Raman spectrum under uniaxial strain conditions, we employed convex and concave molds that have different curve radii ranging from 110 mm to 20 mm, which corresponded to uniaxial strains of up to 0.89% of the tensile strain (for the convex mold) and up to 0.24% of the compressive strain (for the concave mold). The strain-dependent characteristics of the A_g^1, B_{2g}, and A_g^2 modes are shown in Figure 5. While the three different modes showed the same qualitative behavior with respect to the applied strains and exhibited a linear increase (blue shift), the rate of increase was different for each phonon mode. In order to examine the degree of peak shifting, the peak shift as a function of strain was plotted, as shown in Figure 6. The peak shifting values of A_g^1, B_{2g}, and A_g^2 were measured to be 0.86 cm^{-1}/$\Delta\varepsilon$, 0.63 cm^{-1}/$\Delta\varepsilon$, and 0.21 cm^{-1}/$\Delta\varepsilon$, respectively. The out-of-plane A_g^1 mode resulted from the opposing vibrations of the top and bottom P atoms with respect to each other within the same layer. The B_{2g}, and A_g^2 modes were associated with the in-plane vibration of the P atoms in different directions [34,35]. Therefore, it is reasonable to assume that the A_g^1 mode was slightly more sensitive under bending conditions, because the crystal distortion in the vertical direction was more than in the horizontal direction under uniaxial strain. However, these measured slopes were smaller than in the other BP strain studies. It is speculated that the difference derives from the vibrational interaction and light diffraction between BP and the substrate, which was also observed in the other two-dimensional materials [36].

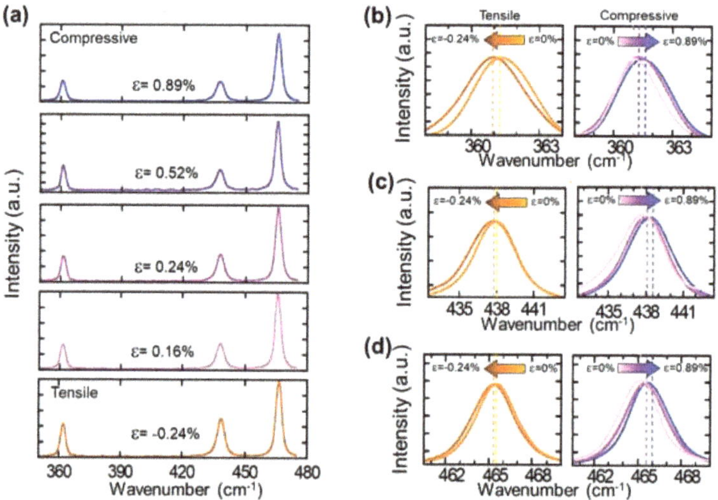

Figure 5. (**a**) Raman spectra of BP measured under uniaxial strains of up to 0.89% of the tensile strain (for the convex mold) and up to 0.24% of the compressive strain. (**b–d**) Magnified Raman spectra of the A_g^1, B_{2g}, and A_g^2 Raman modes as a function of the applied strain.

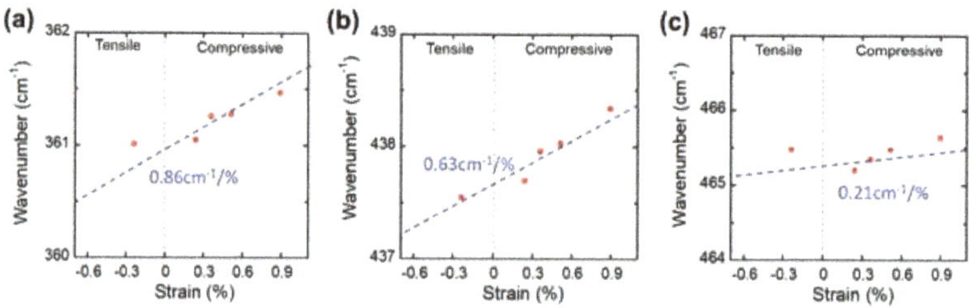

Figure 6. P peaks of (**a**) A_g^1, (**b**) B_{2g}, and (**c**) A_g^2 Raman modes as a function of the applied strain.

4. Conclusions

In summary, we have systematically studied the Raman vibration of BP transferred onto a Ge-coated PDMS substrate, which provided a much higher BP contrast in BP compared to the uncoated substrate and allowed us to investigate much thinner layers of BP directly on a flexible substrate. The Raman spectra taken from several BP layers with different thickness revealed that the clear peak shifting rates for the A_g^1, B_{2g}, and A_g^2 modes were 0.15 cm^{-1}/nm, 0.11 cm^{-1}/nm, and 0.11 cm^{-1}/nm, respectively. Also, the full width at half maximum (FWHM) of all three Raman phonon modes increased as the thickness of the BP layer decreased. Particularly, the peak of the A_g^2 mode increased faster than those of the A_g^1 and B_{2g} phonon modes, indicating that Raman spectra provide a simple way to identify the number of layers of BP. Using this parameter to identify the 2–3-layered BP, a stain–Raman relationship spectrum study was conducted with a maximum uniaxial strain of 0.89%. The peak shifting values of the A_g^1, B_{2g}, and A_g^2 modes by uniaxial strain were measured to be 0.86 cm^{-1}/$\Delta\varepsilon$, 0.63 cm^{-1}/$\Delta\varepsilon$, and 0.21 cm^{-1}/$\Delta\varepsilon$, respectively. Therefore, the phonon mode peak shifting is a good indicator to gauge strain information.

Author Contributions: All authors performed the research. J.-H.S. conceived the idea and designed and managed the research. S.L. and M.N.H. conducted the experiments. S.L. and J.-H.S. wrote the paper.

Funding: This work was supported by the University at Buffalo Innovative Micro-Programs Accelerating Collaboration in Themes (IMPACT) program and the New York State Center of Excellence in Materials Informatics (CMI) program.

Conflicts of Interest: The authors declare no conflict of interest.

References

1. Fei, R.; Yang, L. Strain-engineering the anisotropic electrical conductance of few-layer black phosphorus. *Nano Lett.* **2014**, *14*, 2884–2889. [CrossRef]
2. Guinea, F.; Katsnelson, M.; Geim, A. Energy gaps and a zero-field quantum Hall effect in graphene by strain engineering. *Nat. Phys.* **2010**, *6*, 30–33. [CrossRef]
3. Conley, H.J.; Wang, B.; Ziegler, J.I.; Haglund, R.F., Jr.; Pantelides, S.T.; Bolotin, K.I. Bandgap engineering of strained monolayer and bilayer MoS$_2$. *Nano Lett.* **2013**, *13*, 3626–3630. [CrossRef] [PubMed]
4. Zhou, H.; Seo, J.H.; Paskiewicz, D.M.; Zhu, Y.; Celler, G.K.; Voyles, P.M.; Zhou, W.; Lagally, M.G.; Ma, Z. Fast flexible electronics with strained silicon nanomembranes. *Sci. Rep.* **2013**, *3*, 1291. [CrossRef] [PubMed]
5. Xie, X.; Bai, H.; Shi, G.; Qu, L. Load-tolerant, highly strain-responsive graphene sheets. *J. Mater. Chem.* **2011**, *21*, 2057–2059. [CrossRef]
6. Singh, V.; Joung, D.; Zhai, L.; Das, S.; Khondaker, S.I.; Seal, S. Graphene based materials: Past, present and future. *Prog. Mater. Sci.* **2011**, *56*, 1178–1271. [CrossRef]
7. Wang, Y.; Yang, R.; Shi, Z.; Zhang, L.; Shi, D.; Wang, E.; Zhang, G. Super-elastic graphene ripples for flexible strain sensors. *ACS Nano* **2011**, *5*, 3645–3650. [CrossRef]
8. Bissett, M.A.; Konabe, S.; Okada, S.; Tsuji, M.; Ago, H. Enhanced chemical reactivity of graphene induced by mechanical strain. *ACS Nano* **2013**, *7*, 10335–10343. [CrossRef] [PubMed]
9. Lee, J.E.; Ahn, G.; Shim, J.; Lee, Y.S.; Ryu, S. Optical separation of mechanical strain from charge doping in graphene. *Nat. Commun.* **2012**, *3*, 1024. [CrossRef] [PubMed]
10. Choi, S.-M.; Jhi, S.-H.; Son, Y.-W. Effects of strain on electronic properties of graphene. *Phys. Rev. B* **2010**, *81*, 081407. [CrossRef]
11. Huang, M.; Pascal, T.A.; Kim, H.; Goddard, W.A., III; Greer, J.R. Electronic-mechanical coupling in graphene from in situ nanoindentation experiments and multiscale atomistic simulations. *Nano Lett.* **2011**, *11*, 1241–1246. [CrossRef] [PubMed]
12. Mikael, S.; Seo, J.-H.; Javadi, A.; Gong, S.; Ma, Z. Wrinkled bilayer graphene with wafer scale mechanical strain. *Appl. Phys. Lett.* **2016**, *108*, 183101. [CrossRef]
13. Xia, F.; Wang, H.; Jia, Y. Rediscovering black phosphorus as an anisotropic layered material for optoelectronics and electronics. *Nat. Commun.* **2014**, *5*, 4458. [CrossRef]

14. Ling, X.; Wang, H.; Huang, S.; Xia, F.; Dresselhaus, M.S. The renaissance of black phosphorus. *Proc. Natl. Acad. Sci. USA* **2015**, *112*, 4523–4530. [CrossRef]

15. Morita, A. Semiconducting black phosphorus. *Appl. Phys. A* **1986**, *39*, 227–242. [CrossRef]

16. Qiao, J.; Kong, X.; Hu, Z.-X.; Yang, F.; Ji, W. High-mobility transport anisotropy and linear dichroism in few-layer black phosphorus. *Nat. Commun.* **2014**, *5*, 4475. [CrossRef] [PubMed]

17. Castellanos-Gomez, A. Black phosphorus: Narrow gap, wide applications. *J. Phys. Chem. Lett.* **2015**, *6*, 4280–4291. [CrossRef]

18. Koenig, S.P.; Doganov, R.A.; Schmidt, H.; Neto, A.C.; Oezyilmaz, B. Electric field effect in ultrathin black phosphorus. *Appl. Phys. Lett.* **2014**, *104*, 103106. [CrossRef]

19. Liu, H.; Neal, A.T.; Zhu, Z.; Luo, Z.; Xu, X.; Tománek, D.; Ye, P.D. Phosphorene: An unexplored 2D semiconductor with a high hole mobility. *ACS Nano* **2014**, *8*, 4033–4041. [CrossRef]

20. Li, L.; Yu, Y.; Ye, G.J.; Ge, Q.; Ou, X.; Wu, H.; Feng, D.; Chen, X.H.; Zhang, Y. Black phosphorus field-effect transistors. *Nat. Nanotechnol.* **2014**, *9*, 372. [CrossRef] [PubMed]

21. Wang, H.; Wang, X.; Xia, F.; Wang, L.; Jiang, H.; Xia, Q.; Chin, M.L.; Dubey, M.; Han, S.J. Black phosphorus radio-frequency transistors. *Nano Lett.* **2014**, *14*, 6424–6429. [CrossRef]

22. Engel, M.; Steiner, M.; Avouris, P. Black phosphorus photodetector for multispectral, high-resolution imaging. *Nano Lett.* **2014**, *14*, 6414–6417. [CrossRef] [PubMed]

23. Yuan, H.; Liu, X.; Afshinmanesh, F.; Li, W.; Xu, G.; Sun, J.; Lian, B.; Curto, A.G.; Ye, G.; Hikita, Y.; et al. Polarization-sensitive broadband photodetector using a black phosphorus vertical p–n junction. *Nat. Nanotechnol.* **2015**, *10*, 707–713. [CrossRef] [PubMed]

24. Kim, J.; Baik, S.S.; Ryu, S.H.; Sohn, Y.; Park, S.; Park, B.G.; Denlinger, J.; Yi, Y.; Choi, H.J.; Kim, K.S. Observation of tunable band gap and anisotropic Dirac semimetal state in black phosphorus. *Science* **2015**, *349*, 723–726. [CrossRef] [PubMed]

25. Peng, X.; Wei, Q.; Copple, A. Strain-engineered direct-indirect band gap transition and its mechanism in two-dimensional phosphorene. *Phys. Rev. B* **2014**, *90*, 085402. [CrossRef]

26. Han, X.; Stewart, H.M.; Shevlin, S.A.; Catlow, C.R.A.; Guo, Z.X. Strain and orientation modulated bandgaps and effective masses of phosphorene nanoribbons. *Nano Lett.* **2014**, *14*, 4607–4614. [CrossRef]

27. Zhu, W.; Park, S.; Yogeesh, M.N.; McNicholas, K.M.; Bank, S.R.; Akinwande, D. Black phosphorus flexible thin film transistors at gighertz frequencies. *Nano Lett.* **2016**, *16*, 2301–2306. [CrossRef]

28. Zhu, W.; Yogeesh, M.N.; Yang, S.; Aldave, S.H.; Kim, J.S.; Sonde, S.; Tao, L.; Lu, N.; Akinwande, D. Flexible black phosphorus ambipolar transistors, circuits and AM demodulator. *Nano Lett.* **2015**, *15*, 1883–1890. [CrossRef] [PubMed]

29. Kang, J.; Shin, D.; Bae, S.; Hong, B.H. Graphene transfer: Key for applications. *Nanoscale* **2012**, *4*, 5527–5537. [CrossRef]

30. Kidger, M.J. *Fundamental Optical Design*; SPIE Press: Bellingham, WA, USA, 2002.

31. Mao, N.; Tang, J.; Xie, L.; Wu, J.; Han, B.; Lin, J.; Deng, S.; Ji, W.; Xu, H.; Liu, K.; et al. Optical anisotropy of black phosphorus in the visible regime. *J. Am. Chem. Soc.* **2015**, *28*, 300–305. [CrossRef]

32. Fei, R.; Yang, L. Lattice vibrational modes and Raman scattering spectra of strained phosphorene. *Appl. Phys. Lett.* **2014**, *105*, 083120. [CrossRef]

33. Guo, Z.; Zhang, H.; Lu, S.; Wang, Z.; Tang, S.; Shao, J.; Sun, Z.; Xie, H.; Wang, H.; Yu, X.F.; et al. From black phosphorus to phosphorene: Basic solvent exfoliation, evolution of Raman scattering, and applications to ultrafast photonics. *Adv. Funct. Mater.* **2015**, *25*, 6996–7002. [CrossRef]

34. Ling, X.; Huang, S.; Hasdeo, E.H.; Liang, L.; Parkin, W.M.; Tatsumi, Y.; Nugraha, A.R.; Puretzky, A.A.; Das, P.M.; Sumpter, B.G.; et al. Anisotropic Electron-Photon and Electron-Phonon Interactions in Black Phosphorus. *Nano Lett.* **2016**, *16*, 2260–2267. [CrossRef]

35. Calizo, I.; Bao, W.; Miao, F.; Lau, C.N.; Balandin, A.A. The effect of substrates on the Raman spectrum of graphene: Graphene- on-sapphire and graphene-on-glass. *Appl. Phys. Lett.* **2007**, *91*, 201904. [CrossRef]

36. Wang, Y.Y.; Ni, Z.H.; Yu, T.; Shen, Z.X.; Wang, H.M.; Wu, Y.H.; Chen, W.; Shen Wee, A.T. Raman Studies of Monolayer Graphene: The Substrate Effect. *J. Phys. Chem. C* **2008**, *112*, 10637–10640. [CrossRef]

Article

Direct Exfoliation of Natural SiO₂-Containing Molybdenite in Isopropanol: A Cost Efficient Solution for Large-Scale Production of MoS₂ Nanosheetes

Wenyan Zhao [1], Tao Jiang [1], Yujie Shan [1], Hongrui Ding [2], Junxian Shi [3,*], Haibin Chu [1] and Anhuai Lu [2,*]

1 School of Chemistry and Chemical Engineering, Inner Mongolia University, Hohhot 010021, Inner Mongolia, China; zhaowenyan@imu.edu.cn (W.Z.); Jiang.T@mail.imu.edu.cn (T.J.); shanyj930@163.com (Y.S.); chuhb@imu.edu.cn (H.C.)
2 School of Earth and Space Sciences, Peking University, Beijing 100871, China; DHR@pku.edu.cn
3 School of Ecology and Environment, Inner Mongolia University, Hohhot 010021, Inner Mongolia, China
* Correspondence: 111969116@imu.edu.cn (J.S.); ahlu@pku.edu.cn (A.L.)

Received: 20 September 2018; Accepted: 12 October 2018; Published: 17 October 2018

Abstract: The cost-effective exfoliation of layered materials such as transition metal dichalcogenides into mono- or few- layers is of significant interest for various applications. This paper reports the preparation of few-layered MoS₂ from natural SiO₂-containing molybdenite by exfoliation in isopropanol (IPA) under mild ultrasonic conditions. One- to six-layer MoS₂ nanosheets with dimensions in the range of 50-200 nm are obtained. By contrast, MoS₂ quantum dots along with nanosheets are produced using N-methyl-pyrrolidone (NMP) and an aqueous solution of poly (ethylene glycol)-block-poly (propylene glycol)-block-poly (ethylene glycol) (P123) as exfoliation solutions. Compared with molybdenite, commercial bulk MoS₂ cannot be exfoliated to nanosheets under the same experimental conditions. In the exfoliation process of the mineral, SiO₂ associated in molybdenite plays the role of similar superfine ball milling, which significantly enhances the exfoliation efficiency. This work demonstrates that isopropanol can be used to exfoliate natural molybdenite under mild conditions to produce nanosheets, which facilitates the preparation of highly concentrated MoS₂ dispersions or MoS₂ in powder form due to the volatility of the solvent. Such exfoliated MoS₂ nanosheets exhibit excellent photoconductivity under visible light. Hence, the direct mild exfoliation method of unrefined natural molybdenite provides a solution for low-cost and convenient production of few-layered MoS₂ which is appealing for industrial applications.

Keywords: natural molybdenite; MoS₂ nanosheet; SiO₂; liquid exfoliation; photoelectric properties

---

1. Introduction

Since graphene's initial discovery [1], two-dimensional (2D) transition metal dichalcogenide semiconductors (TMD) have attracted great research interest for their unique electronic and optical characteristics, which are distinctively different from those of their bulk materials [2–6]. MoS₂ is the most popular member of the series, which compensates for the disadvantage caused by the absence of a band gap of graphene, and shows considerable anisotropy due to its large intrinsic band gap, resulting in novel electronic, optical, mechanical, and structural characteristics [7–11]. In many applications, such as batteries, composites, sensors, and catalytic activities, MoS₂ needs to be produced on a large-scale, and preferably at a lower cost.

Different methods to prepare monolayer and few-layered MoS₂ have been developed, including mechanical exfoliation [12], liquid-phase exfoliation [13–15], chemical exfoliation [16,17], and chemical vapor deposition [18,19]. Generally, studies have found that liquid-phase exfoliation has great potential for scalable production of 2D materials. Coleman et al., have indicated that layered materials can be exfoliated into ultrathin-layered 2D nanomaterials in organic solvents by sonication [14]. But for large-scale applications, it has been difficult to achieve high enough concentrations. In many cases, the exfoliating media used with high boiling points are difficult to remove. A method of mixed-solvent containing volatile solvents has been used to exfoliate TMD [20–22]. For example, the mixture of water and ethanol was demonstrated to be an effective solvent for the exfoliation of MoS_2 nanosheets [23]. Nguyen et al. have reported a two-solvent grinding-assisted liquid phase exfoliation of layered MoS_2, avoiding the solvent residue [24]. Ultrasonic treatment is widely used in liquid exfoliation. Ultrasonic force generates acoustic cavitation. The shear forces coming from acoustic cavitation can break the Van der Waals interactions of bulk materials, leading to the exfoliation of 2D materials [14,24], and affecting the structural characteristics of nanoparticles [25]. In many cases, ultrasound probes are utilized to exfoliate 2D materials. Their use can generate violent mechanical driving forces by concentrating energy in a small amount of dispersion liquid. In contrast, an ultrasonic water bath is milder and more cost-effective alternative, which is appealing for samples batch processing. However, it is rarely used in current liquid exfoliation methods.

The intercalation of various intercalates is another useful liquid phase exfoliation method to weaken neighboring layers arising from the interlayer expansion. Peng et al. recently demonstrated a lithium-intercalated single-crystals exfoliation method for 2D TMD nanomaterials in water by only simple manual shaking [26]. However, this chemical exfoliation method partly leads to structural deformation, and is very sensitive to the environmental conditions. To overcome the aforementioned drawback, an alternative approach used grinding/sonication-assisted Li^+ intercalation in simulated sun irradiation conditions, avoiding the use of hazardous liquids such as butyllithium; however, the use of N-methyl-pyrrolidone (NMP) leads to persistent residues on the exfoliated flakes [27]. Great effort has been made to develop new exfoliation technology. Recently, a microcentrifugation surface acoustic wave device was developed to exfoliate MoS_2, which applied an electric field and mechanical shear force for fast and efficient exfoliation. Similarly, the surfactant residues and expensive equipment defeat the purpose [28]. Consequently, it remains challenging to develop liquid exfoliation techniques in more convenient and cost-effective ways to obtain highly-concentrated 2D MoS_2 dispersions that allow separation as precipitates for further application.

In many cases, the preparation of 2D MoS_2 in the lab is based on the liquid exfoliation of MoS_2 powder and single crystals. Savjani et al., [29] and Dong et al., [30] have produced MoS_2 nanosheets by directly exfoliating molybdenite minerals in NMP, and showed that unrefined molybdenite could be an exfoliation source of 2D materials. However, in many cases, the molybdenite powders are purified before exfoliation, and the effects of the impurity constituents on the exfoliation have not been considered. Because natural molybdenite contains quartz, it would thus be of interest to learn whether the quartz in natural molybdenite may play the role of ball mill during the exfoliation of molybdenite, before being separated from the suspension of the MoS_2 nanosheets after exfoliation. Furthermore, we wonder if a solvent whose surface tension does not harshly match with the layered bulk materials can effectively exfoliate natural molybdenite under mild conditions. Relevant studies should be performed to solve these questions.

Here, we prepared 2D MoS_2 nanomaterials by liquid exfoliation of SiO_2-containing molybdenite ores in the volatile solvent, isopropanol (IPA), as well as in NMP, a surfactant aqueous solution under mild water bath ultrasound conditions. The nanomaterials obtained from molybdenite have been compared with the products obtained from commercial MoS_2 powder in the same experimental conditions in terms of exfoliation efficiency, composition, structural features, and photoelectric properties.

2. Materials and Methods

2.1. Materials

Natural molybdenite (NM) was collected and enriched in Dasuji diggings, Zhuozi County, Inner Mongolia. Compositional analysis by X-ray fluorescence (XRF) spectrometry shows that natural molybdenite mainly consisted of Mo, S, SiO_2, Al_2O_3 (Table S1). The exact content of Mo is 49.71wt%, S is 27.65wt%, and SiO_2 is 14.15wt%, which was determined using an inductively-coupled plasma-mass (ICP-MS) spectrometer. The NM was ball-milled (XGB planetary mill, 100 stainless steel balls of 6 mm diameter, rotation speed 500 rpm) for 1 h and sifted to obtain mineral powder with particle sizes of <45 μm. The scanning electron microscope (SEM) images are shown in Figure S1a,b. The X-ray diffraction (XRD) patterns prove that the NM was mainly formed from 2H MoS_2 and quartz-phase SiO_2 (Figure S2). Energy dispersive spectrometry (EDS) shows that the SiO_2 and MoS_2 are mixed uniformly on the micro-scale (Figure S3). Commercially-available MoS_2 (CM, 99%, average 2 μm) was bought from Sigma-Aldrich (Figure S1c,d). NMP, P123, and IPA were purchased from J&K Chemicals (Beijing, China).

Preparation of control samples of NM: SiO_2 associated in NM was removed by HF method; 2 g of NM was added to 20 mL mixed acid solution (16 mL 40%HF and 4 mL H_2SO_4) in 50 mL Teflon crucible, and maintained under magnetic stirring for 3 h at 80 °C; after reaction, the product was filtrated, washed with Milli-Q water several times to neutral, then dried at 60 °C in oven.

2.2. Exfoliation Process

0.5 g of NM or CM was added to 50 mL of exfoliation solution (IPA, NMP or P123 aqueous solutions) in a 100 mL glass vial. The mixture was batch sonicated for 16 h in an ultrasonic water bath with a frequency of 40 kHz (volume 15 L, power 400 W), taking care to have the water level higher than the level of the suspension. The ultrasonic temperature was controlled below 50 °C. The resulting dispersion was centrifuged for 15 min at 10,000 rpm to remove the very thick nanosheets and excess impurity. The supernatants were collected by pipette. The products derived from NM or CM were labeled as NM-X or CM-X respectively, where X is the exfoliation solvent.

2.3. Characterization

Compositional analysis of NM was achieved using a Rigaku ZSX Primus II XRF spectrometer and a Thermo X Series II ICP-MS spectrometer. The UV-vis (Ultraviolet-visible) spectra were performed by a Hitachi UV-3900 spectrometer using 1.0 cm quartz cuvette. Accurate dilutions of the dispersions were produced to obtain suitable UV-vis spectra. The XRD pattern measurements were carried out using a PANalytical X'Pert Pro diffractometer with Cu Kα radiation at 0.15406 nm (equipped with monochromator). Raman spectra were collected by a MicroRaman spectrometer (Renishaw in Via Reflex, Gloucestershire, UK), equipped with a 532 nm DPSS laser and a 2400 lines/mm grating. For the specimen preparation, the suspension was drop-casted onto a silicon substrate (wafer) and heated to 50 °C to form an opaque film. SEM imaging was performed using Hitachi S4800. Energy dispersive X-ray analyses were performed using Bruker QUANTAX 200 energy dispersive spectrometer. The transmission electron microscopy (TEM) images were measured using a FEI Tecnai G2F20S-TWIN. Samples for TEM imaging were prepared by drop-casting the 2D MoS_2 dispersion onto 200 mesh lacey carbon-coated copper grid and dried before analysis. Nanosheets thickness measurements were carried out on a Bruker atomic force microscopy (AFM) (Dimension Icon, Bruker, Billerica, MA, USA) operating in ScanAsyst mode. A silicon nitride probe (type SCANASYST-AIR, Bruker, Billerica, MA, USA) with a curvature radius of 2 nm was used for AFM measurements. Its specifications include a force constant of approximately 0.4 N/m, and a resonant frequency of 70 kHz. Samples for AFM were prepared by diluting the samples in ethanol to a concentration in the range of 100 μg/mL, drop-casting the diluted suspension onto a clean silicon wafer piece, and heating to evaporate the solvent.

2.4. Photoelectrochemical Measurement

The working electrodes were prepared on glassy carbon (GC) electrodes, which were dip-coated with nano MoS_2 dispersions (loading 18 µg/cm^2). The electrochemical measurements were tested in 0.5 M Na_2SO_4 with a conventional three-electrode system. Pt plate and saturated Ag/AgCl were used as counter electrode and reference electrode, respectively. The photocurrents were measured on a CHI760E Chenhua electrochemical workstation. The visible light irradiation source was obtained by a 300 W Xe lamp (Beijing Trustech, Beijing, China, PLS-SXE300c) with a 420 nm cut-off filter.

3. Results and Discussion

Nano MoS_2 dispersions were prepared from NM or CM using a simple bath sonication technique in IPA, NMP, or aqueous solutions of surfactant P123. After bath sonication for 16 h, suspension of different colours were obtained and labeled as NM-NMP, NM-P123, NM-IPA, CM-NMP, CM-P123, and CM-IPA respectively (Figure 1 inset). Products from NM are dark green, CM-NMP is golden yellow, CM-P123 is yellow-green, and CM-IPA is colourless and transparent. Except for CM-IPA, the Tyndall effect was observed, which indicates that the suspension was a colloidal dispersion. UV-vis spectra of the exfoliated MoS_2 are shown in Figure 1. In the samples derived from NM, the characteristic absorption bands of MoS_2 nanosheets at 390nm (A), 450nm (B), 610nm (C), and 670 nm (D) are clearly observed [31–33]. The excitonic peaks at 610 nm and 670 nm arise from the K point of the Brillouin zone. The peaks at ~390 nm and ~450 nm can be attributed to the direct transition from the deep valence band to the conduction band [34,35]. But in the samples derived from CM, absorption bands in the visible light region are very weak. These results from UV-vis absorption only confirm that NM can be exfoliated to produce MoS_2 nanosheets in three solvents, but CM cannot be exfoliated to obtain nanosheets under mild ultrasonic conditions.

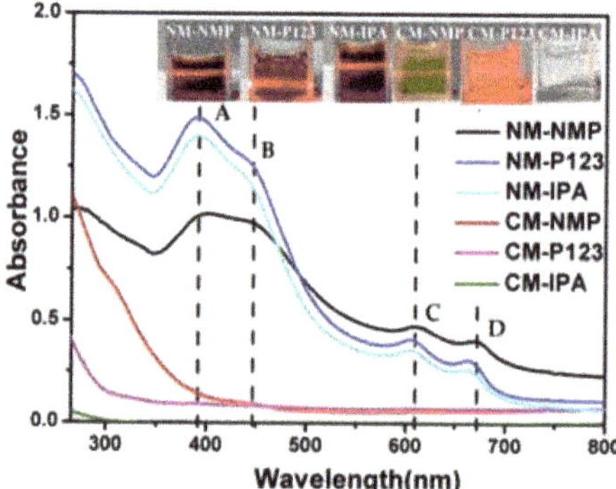

Figure 1. Ultraviolet-visible absorption spectra of products exfoliated in various solvents from natural molybdenite (NM) and commercial MoS_2 (CM). The Photographs of different products with a red laser crossed through is shown in the inset.

Further characterization was studied using the TEM images of exfoliated MoS_2 for NM-NMP, CM-NMP, and NM-IPA. TEM images of NM-NMP are shown in Figure 2a,b,c. The transparency of the nanosheets confirms that the MoS_2 sheets are very thin. Clear lattice fringes with a separation of 0.2725 nm are observed for (100) planes of 2H-MoS_2 in Figure 2a. The selected area electron diffraction (SAED) pattern of MoS_2 nanosheets shown in the left inset of Figure 2a reveals the highly crystalline nature of the sheets with a hexagonal diffraction pattern. The diffraction spots shown by the SAED correspond to the (100), (110), (200) crystal faces of the 2H-MoS_2. Figure 2b shows several erected nanosheets. The nanosheets have only a few molecular layers, with interlayer spacing of 0.623 nm corresponding to the (002) crystal plane; the atomic arrangements are shown in Figure 2b inset. In addition, MoS_2 quantum dots around the nanosheets are clearly observed in Figure 2c. Figure 2d shows the image of CM-NMP, in which quantum dots of ~5nm are observed. The left inset of Figure 2d shows SAED pattern of quantum dots which correspond to the (100), (105) crystal faces of hexagonal system, and the inset at the top right corner shows the hexagonal arrangement of 2H-MoS$_2$. The above results indicate that MoS$_2$ quantum dots, along with nanosheets, were obtained by exfoliating NM; meanwhile, only quantum dots were produced from CM powder in NMP. TEM images of NM-IPA are shown in Figure 2e,f. MoS$_2$ nanosheets with lateral size ~100 nm are observed (Figure 2e), and the layer structure of the edge warpage can be clearly shown in Figure 2f (arrow indication). Lattice fringes with a separation of 0.2745 nm are responsible for (100) planes of 2H-MoS$_2$, and the inset at the top left corner shows the hexagonal arrangement of 2H-MoS$_2$. MoS$_2$ nanosheets are the only products observed in NM-IPA. The majority of the exfoliated MoS$_2$ nanosheets are close to 1–6 layers thickness. More images for the other products obtained, such as NM-P123 and CM-P123, can be found in Figure S4. Quantum dots spreading nanosheets of MoS$_2$ were observed in NM-P123 (Figure S4a,b), and amounts of quantum dots and a few nanosheets were observed in CM-P123 (Figure S4c,d). To further ascertain the morphology and thickness of exfoliated MoS$_2$, AFM was carried out. Figure 3 shows scan AFM images in ScanAsyst mode of MoS$_2$ nanosheets in NM-NMP and NM-IPA samples. Figure 3a,b reveals that the thickness of nanosheets was less than 6 nm both in NM-NMP and NM-IPA. Given that the thickness of a MoS$_2$ monolay is ~1 nm [14,23,36], this suggests that the obtained MoS$_2$ nanosheets contain ~1 to 6 layers, which agrees with the TEM results.

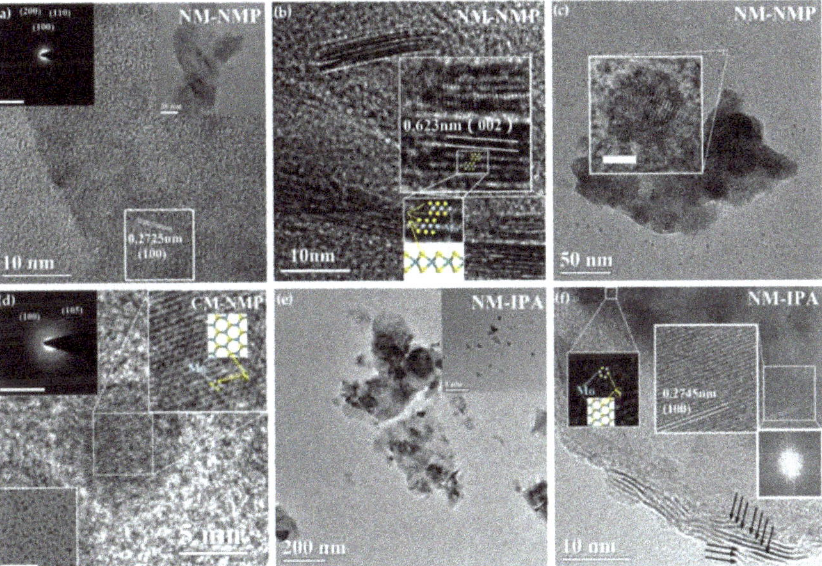

Figure 2. The high-resolution transmission electron microscopy (HRTEM) images of products exfoliated in N-methyl-pyrrolidone (NMP) from NM (**a, b, c**), CM (**d**), and products exfoliated in isopropanol (IPA) from NM (**e, f**). Insets in (**a**): a selected area electron diffraction (SAED) pattern, scale bar = 5 nm^{-1}, a low resolution image, scale bar = 20 nm; insets in (**b, f**): magnification of the selected area and schematic diagram of atomic arrangement; inset in (**c**): a quantum dot image, scale bar =2 nm; insets in (**d**): a SAED pattern, scale bar = 10 nm^{-1}, a low resolution image, scale bar = 50 nm, magnification of the selected area and schematic diagram of atomic arrangement; inset in (**e**): a low resolution image, scale bar = 1 μm.

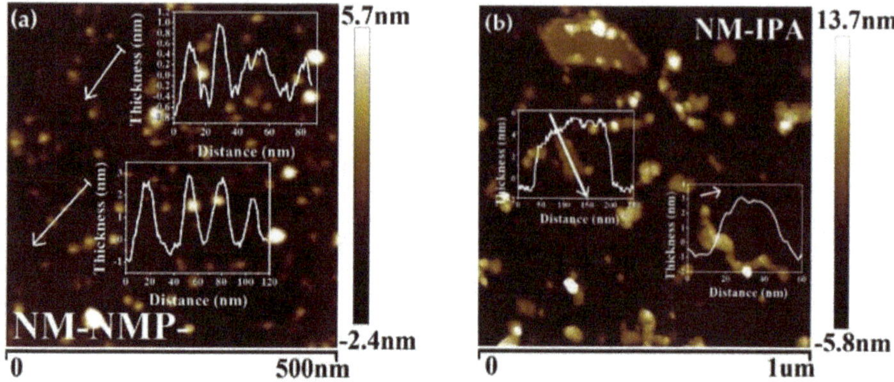

Figure 3. Atomic force microscopy (AFM) images of products exfoliated from NM in NMP (**a**) and in IPA (**b**). The insets show the height profiles along with the arrows. AFM images and height profile confirming the thickness of <6 nm.

To further confirm the phase, XRD and Raman spectroscopy were performed (Figure 4). Due to the presence of polymeric surfactant P123, we could not use AFM, XRD, or Raman to analyze product NM-P123 unless the samples were treated by washing or calcining, which could change the samples (Figure S5 and S6). The appearance of (002) reflection in the XRD patterns of NM-NMP and NM-IPA indicates the presence of MoS_2 nanosheets derived from NM with good crystallinity (Figure 4a). Meanwhile, XRD pattern of CM-NMP exhibiting no apparent reflection also indicates that CM-NMP mainly consists of quantum dots. Raman spectra are shown in Figure 4b. The characteristic shifts near 380.5 cm^{-1} and 406.5 cm^{-1} for bulk NM, NM-NMP, and NM-IPA respectively correspond to E^1_{2g} and A_{1g} vibrations modes of Mo-S bands in 2H MoS_2 [37]. After exfoliation, the E^1_{2g} and A_{1g} peak shift, and the shift difference between E^1_{2g} and A_{1g}, is generally only changed with the layer number of molecules. So, the reduction of the shift difference between the two peaks can be considered to be an indicator of the layer number of MoS_2 nanosheets [38]. Compared to 26.0 cm^{-1} shift difference of the two characteristic peaks for bulk NM, the shift difference reduces to 24.7 cm^{-1} for NM-NMP and 24.1 cm^{-1} for NM-IPA respectively. On the basis of information provided by Li et al., [37], this result indicates that the thickness of MoS_2 nanosheets exfoliated from NM was mostly less than 6 layers, which is consistent with the measurements of TEM and AFM.

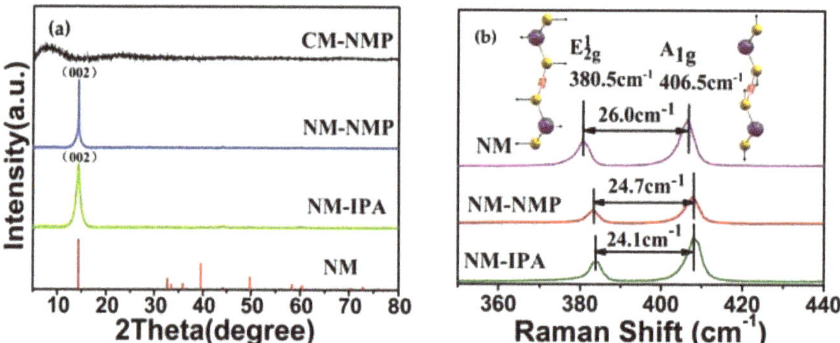

Figure 4. (**a**) X-ray diffraction (XRD) patterns of NM-NMP, NM-IPA, CM-NMP and bulk NM. (**b**) Raman spectra of bulk NM, NM-NMP and NM-IPA.

MoS$_2$ concentration in suspension was measured by atomic absorption spectroscopy. The concentrations of dispersed nano MoS$_2$ exfoliated from NM are 596, 485, and 252 mg/L for P123, NMP, and IPA solution respectively. Compared to the concentrations of products exfoliated from CM, which are listed in Table 1, the concentration of MoS$_2$ derived from NM is relatively high. Furthermore, nanosheets are more easily obtained from NM. The surprise result raises a question: why can NM with lower purity and larger particle sizes compared with CM be exfoliated to produce more nanosheets? In many exfoliation methods, MoS$_2$ nanosheets were obtained using an ultrasound probe as a violent mechanical driving force because it concentrates more energy [14,36,39]. In our case, due to the dispersive energy of the ultrasonic water bath, MoS$_2$ quantum dots separated from the crystal defects were the only product from CM [40]. However, MoS$_2$ nanosheets were obtained under the same mild ultrasound conditions from NM. NM contains small amounts of impurities compared with CM, mainly SiO$_2$ (Table S1). Thus, it seems reasonable to hypothesize that the quartz phase SiO$_2$ in the NM may cause collisions with MoS$_2$ during bath sonication, leading to cleavage of bulk layers into ultrathin sheets, which can play the role of superfine ball milling. Furthermore, compared with the added ball milling medium (e.g., ZrO$_2$) [39], associated quartz in the NM bonds with MoS$_2$ more closely and uniformly, and the particle size of that is much smaller, thereby causing more effective stripping. After exfoliation, most of the SiO$_2$ was removed from dispersion by centrifugation, as observed in the EDS. Quantitative analysis shows that Si content in exfoliated nanosheets is 0.32wt%, which is much lower than that in natural molybdenite (5.27wt%). EDS analysis shows Mo : S to be 1 : 2 in all case, and reveals that the nano MoS$_2$ prepared from NM does not contain more non-negligible impurity contaminations than that obtained from CM.

Table 1. The concentrations and morphologies of MoS$_2$ exfoliated.

	NM-NMP	NM-P123	NM-IPA	CM-NMP	CM-P123	CM-IPA
Concentration of MoS$_2$ (mg/L)	485	596	252	239	485	—[1]
Morphology of MoS$_2$	Nanosheets & quantum dots	Nanosheets & quantum dots	Nanosheets	quantum dots	Nanosheets & quantum dots	—[1]

[1] No detection.

To test the aforementioned hypotheses, we prepared control samples of NM which were purified to remove SiO$_2$ by the HF method, and compared them with the initial NM as the source powders of exfoliation in IPA. After HF treatment, the SiO$_2$ content in molybdenite reduces from 11% to 0.65%, and the absorbance intensity of exfoliation products at ca. 670 nm plummets, as shown in Figure S7, indicating that the yield of the nanosheets from NM with SiO$_2$ removed is considerably lower than that from initial NM. Therefore, this result supports our argument about the associated SiO$_2$ dependent exfoliation mechanism of the mineral, as shown in Scheme 1.

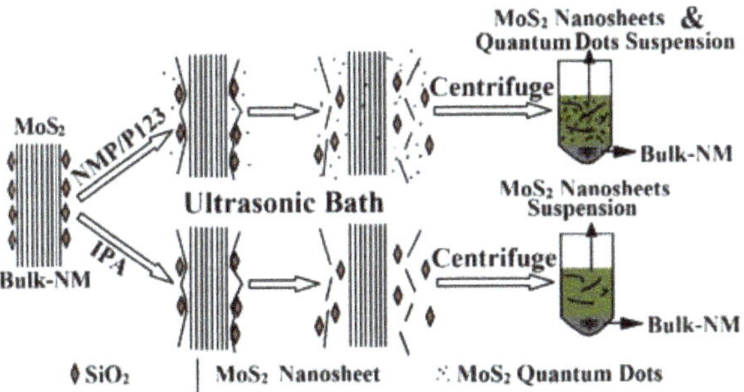

Scheme 1. Exfoliation mechanism of natural SiO$_2$-associated molybdenite in IPA, NMP, and P123 aqueous solution, under mild ultrasonic condition.

In addition, the exfoliation efficiency varies greatly with the solvents, resulting in a different morphology and yield of exfoliated nano-MoS_2. During the wet grinding process, the choice of grinding solvent has a significant influence on the final product [24]. According to the principle of matching surface energy of solvents [41], NMP and 10 wt% P123 with a surface tension close to 40 mN/m are good solvents for exfoliating bulk MoS_2; therefore, MoS_2 quantum dots are easy to obtain by water bath sonication in NMP or P123. As shown in the height profile of AFM (Figure 3), NMP produces thinner nanosheets than IPA. Studies have suggested that a close match of the surface energy between the solvent and the layered material leads to the exfoliation of flakes with increased aspect ratios [42]. Thinner flakes are more liable to breakdown and produce quantum dots by ultrasonic vibration resulting from the ultrasonic water bath. Because IPA surface tension is 20.80 mN/m, far away from 40 mN/m, nano MoS_2 is not successfully prepared in IPA from CM by mild water bath sonication; however, it can be prepared from NM. Due to the surface tension of IPA not being matched, the quantum dots cannot be exfoliated off from the crystal defects, but MoS_2 nanosheets have been achieved in IPA via a mineral-associated quartz milling process. Though the concentrations of dispersed MoS_2 nanosheets in IPA are relatively low, the highly concentrated MoS_2 nanosheets dispersion or its powder form can be obtained due to the volatility of IPA, and the solvent can be recovered for recycling exfoliation to improve yields. Therefore, aggregation of MoS_2 nanosheets caused by extracting exfoliated MoS_2 from NMP via heating evaporation at high temperature can be avoided. Thereafter, the obtained powder from IPA is convenient for characterization and preparation of samples in applications avoiding the effect of a solvent, which could be widely applied.

Two-dimensional MoS_2 has shown prior promise for various energy-related applications because of its intrinsic band-gap structure. To confirm the photoelectric properties of our exfoliated MoS_2, simple light-to-electric conversion experiments were conducted. The time-dependent photocurrent of exfoliated nano MoS_2 was measured at 0.6 V bias voltage under alternating darkness and visible light conditions. As shown in Figure 5, all the samples exhibited repeatable and stable responses to illumination. In contrast, the photocurrents of NM-IPA and NM-NMP are higher than those of bulk NM and CM, and the highest ΔI is generated by NM-IPA, which is around 9-fold higher than that of bare GC. The remarkable increase in photocurrent density results from the sensitivity of exfoliated MoS_2 nanosheets to visible light and the efficient separation of photo-generated electron-hole pair. Unexpectedly, a lower photocurrent response of CM-NMP was observed, which is close to that of bare GC. A higher photocurrent response of MoS_2 nanosheets (NM-IPA) in comparison to their bulk structure gives rise to the change in band structure from indirect to direct bandgap [43,44]. The remarkable decrease in the photocurrent density of CM-NMP (MoS_2 quantum dots) was observed because of its insensitivity to visible light and the quantum confinement effect.

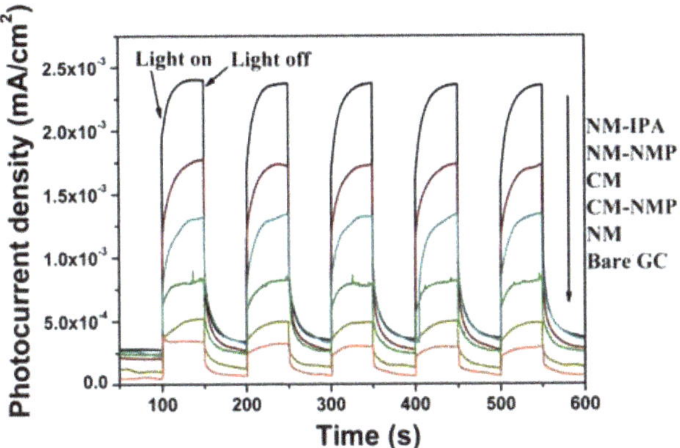

Figure 5. The transient on/off photocurrent responses vs. time plotted for NM-IPA, NM-NMP, CM-NMP, NM, CM, and bare GC in 0.01 M Na_2SO_4 solution under visible-light.

4. Conclusions

Few-layered MoS_2 was prepared from natural SiO_2-containing molybdenite by exfoliation in IPA under mild ultrasonic conditions. The exfoliation products from NM and CM in different exfoliation solvents including IPA, NMP, and aqueous solutions of P123 were compared. The results show that the yield of nano MoS_2 exfoliated from the molybdenite is higher than that from CM powder. The choice of exfoliation solvent plays a crucial role. IPA can successfully exfoliate molybdenite to produce nanosheets, but cannot exfoliate commercial MoS_2 powder. NMP and aqueous solutions of P123 can exfoliate NM to MoS_2 nanosheets and quantum dots, and exfoliate CM to quantum dots. It was found that the MoS_2 nanosheets produced from NM in IPA exhibit greater photocurrent density under visible light. The quartz phase associated in molybdenite is a major contributor to the effective exfoliation, and plays the role of superfine ball milling.

This work highlights that unrefined natural SiO_2-containing NM can be used as an exfoliation source of 2D MoS_2; the ability to use volatile IPA to exfoliate it will enable recirculated exfoliation to produce MoS_2 nanosheets with higher yields, for cost–effective and large-scale industrial applications. However, in the exfoliation process of minerals, the solvent has an influence on the morphology of the product, which is poorly understood at present. Further research may lead to proper evaluations of the effects of different solvents on exfoliation, which help us to fully elucidate the underlying exfoliation mechanisms.

Supplementary Materials: The following are available online at http://www.mdpi.com/2079-4991/8/10/843/s1, Figure S1: SEM images of natural molybdenite (a), sifted natural molybdenite after ball-milled (b) and commercial MoS_2 (c, d); Figure S2: XRD patterns of natural molybdenite and commercial MoS_2; Figure S3: Elemental analysis of natural molybdenite using EDS shows the SiO_2 and MoS_2 are mixed uniformly in the micro-scale; Figure S4: TEM images of products exfoliated in P123 aqueous solution from natural molybdenite (a, b), commercial MoS_2 (c, d); Figure S5: (a) XRD patterns of products exfoliated in P123 from natural molybdenite and commercial MoS_2, (b) TEM images of NM-P123 powder-sample obtained by washing with deioned water several times to remove P123 after centrifuging at 1500rpm for 45min; Figure S6: (a) Raman spectrum of NM-P123 powder-sample obtained by calcining at 450°C for 2h, (b) TGA curves of NM, CM, NM-IPA-Powder and P123; Figure S7: UV-Vis absorption spectra stack plot of MoS_2 dispersions obtained from the initial NM and NM removed SiO_2, Table S1: Content of natural molybdenite by component analysis using XRF.

Author Contributions: Conceptualization, W.Z. and J.S.; methodology, T.J., W.Z; formal analysis, T.J., Y.S. and H.D.; resources, J.S. and W.Z.; writing—original draft preparation, W.Z., T.J. and Y.S.; writing—review and editing, H.C. and J.S.; funding acquisition, A.L., J.S. and W.Z.; supervision, A.L.; project administration, A.L.

Funding: This research was funded by the National Basic Research Program of China (2014CB846001), National Natural Science Foundation of China (21167008) and Inner Mongolia Natural Science Foundation (2015MS0203).

Conflicts of Interest: The authors declare no conflict of interest.

References

1. Novoselov, K.S.; Geim, A.K.; Morozov, S.V.; Jiang, D.; Zhang, Y.; Dubonos, S.V.; Grigorieva, I.V.; Firsov, A.A. Electric field effect in atomically thin carbon films. *Science* **2004**, *306*, 666–669. [CrossRef] [PubMed]
2. Tan, C.; Zhang, H. Two-dimensional transition metal dichalcogenide nanosheet-based composites. *Chem. Soc. Rev.* **2015**, *44*, 2713–2731. [CrossRef] [PubMed]
3. Zhang, H. Ultrathin two-dimensional nanomaterials. *ACS Nano* **2015**, *9*, 9451–9469. [CrossRef] [PubMed]
4. Zhang, X.; Lai, Z.C.; Tan, C.L.; Zhang, H. Solution-processed two-dimensional MoS_2 nanosheets: preparation, hybridization, and application. *Angew. Chem. Int. Ed.* **2016**, *55*, 8816–8838. [CrossRef] [PubMed]
5. Zhang, M.; Zhu, Y.; Wang, X.; Feng, Q.; Qiao, S.; Wen, W.; Chen, Y.; Cui, M.; Zhang, J.; Cai, C.; et al. Controlled synthesis of ZrS_2 monolayer and few layers on hexagonal boron nitride. *J. Am. Chem. Soc.* **2015**, *137*, 7051–7054. [CrossRef] [PubMed]
6. Xia, D.; Gong, F.; Pei, X.D.; Wang, W.B.; Li, H.; Zeng, W.; Wu, M.Q.; Papavassiliou, D.V. Molybdenum and tungsten disulfides-based nanocomposite films for energy storage and conversion: A review. *Chem. Eng. J.* **2018**, *348*, 908–928. [CrossRef]
7. Nguyen, T.P.; Sohn, W.; Oh, J.H.; Jang, H.W.; Kim, S.Y. Size-dependent properties of two-dimensional MoS_2 and WS_2. *J. Phys. Chem. C* **2016**, *120*, 10078–10085. [CrossRef]
8. Rao, C.N.R.; Gopalakrishnan, K.; Maitra, U. Comparative study of potential applications of graphene, MoS_2, and other two-dimensional materials in energy devices, sensors, and related areas. *ACS Appl. Mater. Interfaces* **2015**, *7*, 7809–7832. [CrossRef] [PubMed]
9. Huang, Y.X.; Guo, J.H.; Kang, Y.J.; Ai, Y.; Li, C.M. Two dimensional atomically thin MoS_2 nanosheets and their sensing applications. *Nanoscale* **2015**, *7*, 19358–19376. [CrossRef] [PubMed]
10. Xia, S.S.; Wang, Y.R.; Liu, Y.; Wu, C.H.; Wu, M.H.; Zhang, H.J. Ultrathin MoS_2 nanosheets tightly anchoring onto nitrogen-doped graphene for enhanced lithium storage properties. *Chem. Eng. J.* **2018**, *332*, 431–439. [CrossRef]
11. Jayabal, S.; Saranya, G.; Wu, J.; Liu, Y.Q.; Geng, D.S.; Meng, X.B. Understanding the high-electrocatalytic performance of two-dimensional MoS_2 nanosheets and their composite materials. *J. Mater. Chem. A* **2017**, *5*, 24540–24563. [CrossRef]
12. Yin, Z.; Li, H.; Li, H.; Jiang, L.; Shi, Y.; Sun, Y.; Lu, G.; Zhang, Q.; Chen, X.; Zhang, H. Single-layer MoS_2 phototransistors. *ACS Nano* **2011**, *6*, 74–80. [CrossRef] [PubMed]
13. Cunningham, G.; Lotya, M.; Cucinotta, C.S.; Sanvito, S.; Bergin, S.D.; Menzel, R.; Shaffer, M.S.P.; Coleman, J.N. Solvent exfoliation of transition metal dichalcogenides: dispersibility of exfoliated nanosheets varies only weakly between compounds. *ACS Nano* **2012**, *6*, 3468–3480. [CrossRef] [PubMed]
14. Coleman, J.N.; Lotya, M.; O'Neill, A.; Bergin, S.D.; King, P.J.; Khan, U.; Young, K.; Gaucher, A.; De, S.; Smith, R.J.; et al. Two-dimensional nanosheets produced by liquid exfoliation of layered Materials. *Science* **2011**, *331*, 568–571. [CrossRef] [PubMed]
15. Liu, Y.; He, X.; Hanlon, D.; Harvey, A.; Coleman, J.N.; Li, Y. Liquid phase exfoliated MoS_2 nanosheets percolated with carbon nanotubes for high volumetric/areal capacity sodium-ion batteries. *ACS Nano* **2016**, *10*, 8821–8828. [CrossRef] [PubMed]
16. Zeng, Z.; Sun, T.; Zhu, J.; Huang, X.; Yin, Z.; Lu, G.; Fan, Z.; Yan, Q.; Hng, H.H.; Zhang, H. An effective method for the fabrication of few-layer-thick inorganic nanosheets. *Angew. Chem. Int. Ed.* **2012**, *51*, 9052–9056. [CrossRef] [PubMed]
17. Paredes, J.I.; Munuera, J.M.; Villar-Rodil, S.; Guardia, L.; Ayán-Varela, M.; Pagán, A.; Aznar-Cervantes, S.D.; Cenis, J.L.; Martínez-Alonso, A.; Tascon, J.M. Impact of covalent functionalization on the aqueous processability, Catalytic Activity, and Biocompatibility of Chemically Exfoliated MoS_2 Nanosheets. *ACS Appl. Mater. Interfaces* **2016**, *8*, 27974–27986. [CrossRef] [PubMed]
18. Lee, Y.H.; Zhang, X.Q.; Zhang, W.; Chang, M.T.; Lin, C.T.; Chang, K.D.; Yu, Y.C.; Wang, J.T.W.; Chang, C.S.; Li, L.J.; et al. Synthesis of large-area MoS_2 atomic layers with chemical vapor deposition. *Adv. Mater.* **2012**, *24*, 2320–2325. [CrossRef] [PubMed]
19. Najmaei, S.; Liu, Z.; Zhou, W.; Zou, X.; Shi, G.; Lei, S.; Yakobson, B.I.; Idrobo, J.C.; Ajayan, P.M.; Lou, J. Vapour phase growth and grain boundary structure of molybdenum disulphide atomic layers. *Nat. Mater.* **2013**, *12*, 754–759. [CrossRef] [PubMed]

20. Halim, U.; Zheng, C.R.; Chen, Y.; Lin, Z.; Jiang, S.; Cheng, R.; Huang, Y.; Duan, X. A rational design of cosolvent exfoliation of layered materials by directly probing liquid-solid interaction. *Nat. Commun.* **2013**, *4*, 3213–3219. [CrossRef] [PubMed]

21. Carey, B.J.; Daeneke, T.; Nguyen, E.P.; Wang, Y.; Ou, J.Z.; Zhuiykov, S.; Kalantar-zadeh, K. Two solvent grinding sonication method for the synthesis of two-dimensional tungsten disulphide flakes. *Chem. Commun.* **2015**, *51*, 3770–3773. [CrossRef] [PubMed]

22. Shen, J.F.; Wu, J.J.; Wang, M.; Dong, P.; Xu, J.X.; Li, X.G.; Zhang, X.; Yuan, J.H.; Wang, X.F.; Ye, M.X.; et al. Surface tension components based selection of cosolvents for efficient liquid phase exfoliation of 2D materials. *Small* **2016**, *12*, 2741–2749. [CrossRef] [PubMed]

23. Zhou, K.G.; Mao, N.N.; Wang, H.X.; Peng, Y.; Zhang, H.L. A mixed-solvent strategy for efficient exfoliation of inorganic graphene analogues. *Angew. Chem. Int. Ed.* **2011**, *50*, 10839–10842. [CrossRef] [PubMed]

24. Nguyen, E.P.; Carey, B.J.; Daeneke, T.; Ou, J.Z.; Latham, K.; Zhuiykov, S.; Kalantar-zadeh, K. Investigation of two-solvent grinding-assisted liquid phase exfoliation of layered MoS_2. *Chem. Mater.* **2015**, *27*, 53–59. [CrossRef]

25. Arrigo, R.; Teresi, R.; Gambarotti, C.; Parisi, F.; Lazzara, G.; Dintcheva, N.T. Sonication-induced modification of carbon nanotubes: effect on the rheological and thermo-oxidative behaviour of polymer-based nanocomposites. *Materials* **2018**, *11*, 383. [CrossRef] [PubMed]

26. Peng, J.; Wu, J.J.; Li, X.T.; Zhou, Y.; Yu, Z.; Guo, Y.Q.; Wu, J.C.; Lin, Y.; Li, Z.J.; Wu, X.J.; et al. Very large-sized transition metal dichalcogenides monolayers from fast exfoliation by manual shaking. *J. Am. Chem. Soc.* **2017**, *139*, 9019–9025. [CrossRef] [PubMed]

27. Wang, Y.; Carey, B.J.; Zhang, W.; Chrimes, A.F.; Chen, L.; Kalantar-zadeh, K.; Ou, J.Z.; Daenekem, T. Intercalated 2D MoS_2 utilizing a simulated sun assisted process: reducing the HER overpotential. *J. Phys. Chem. C* **2016**, *120*, 2447–2455. [CrossRef]

28. Mohiuddin, M.; Wang, Y.; Zavabeti, A.; Syed, N.; Datta, R.S.; Ahmed, H.; Daeneke, T.; Russo, S.P.; Rezk, A.R.; Yeo, L.Y.; et al. Liquid phase acoustic wave exfoliation of layered MoS_2: critical impact of electric field in efficiency. *Chem. Mater.* **2018**, *30*, 5593–5601. [CrossRef]

29. Savjani, N.; Lewis, E.A.; Pattrick, R.A.D.; Haigh, S.J.; O'Brien, P. MoS_2 nanosheet production by the direct exfoliation of molybdenite minerals from several type-localities. *RSC Advances* **2014**, *4*, 35609–35613. [CrossRef]

30. Dong, H.N.; Chen, D.L.; Wang, K.; Zhang, R. High-yield preparation and electrochemical properties of few-Layer MoS_2 nanosheets by exfoliating natural molybdenite powders directly via a coupled ultrasonication-milling process, *Nanoscale Res. Lett.* **2016**, *11*, 409. [CrossRef] [PubMed]

31. Wang, T.; Liu, L.; Zhu, Z.; Papakonstantinou, P.; Hu, J.; Liu, H.; Li, M. Enhanced electrocatalytic activity for hydrogen evolution reaction from self-assembled monodispersed molybdenum sulfide nanoparticles on an Au electrode. *Energy Environ. Sci.* **2013**, *6*, 625–633. [CrossRef]

32. Chikan, V.; Kelley, D.F. Size-dependent spectroscopy of MoS_2 nanoclusters. *J. Phys. Chem. B* **2002**, *106*, 3794–3804. [CrossRef]

33. Wilcoxon, J.P.; Newcomer, P.P.; Samara, G.A. Synthesis and optical properties of MoS_2 and isomorphous nanoclusters in the quantum confinement regime. *J. Appl. Phys.* **1997**, *81*, 7934–7944. [CrossRef]

34. Wilson, J.A.; Yoffe, A.D. The transition metal dichalcogenides discussion and interpretation of the observed optical, electrical and structural properties. *Adv. Phys.* **1969**, *18*, 193–335. [CrossRef]

35. Wilcoxon, J.P.; Samara, G.A. Strong quantum-size effects in a layered semiconductor: MoS_2 nanoclusters. *Phys. Rev. B* **1995**, *51*, 7299–7302. [CrossRef]

36. Smith, R.J.; King, P.J.; Lotya, M.; Wirtz, C.; Khan, U.; De, S.; O'Neill, A.; Duesberg, G.S.; Grunlan, J.C.; Moriarty, G.; et al. Large-scale exfoliation of inorganic layered compounds in aqueous surfactant solutions. *Adv. Mater.* **2011**, *23*, 3944–3948. [CrossRef] [PubMed]

37. Li, H.; Zhang, Q.; Yap, C.C.R.; Tay, B.K.; Edwin, T.H.T.; Olivier, A.; Baillargeat, D. From bulk to monolayer MoS_2: evolution of Raman scattering. *Adv. Funct. Mater.* **2012**, *22*, 1385–1390. [CrossRef]

38. Li, S.L.; Miyazaki, H.; Song, H.; Kuramochi, H.; Nakaharai, S.; Tsukagoshi, K. Quantitative Raman spectrum and reliable thickness identification for atomic layers on insulating substrates. *ACS Nano* **2012**, *6*, 7381–7388. [CrossRef] [PubMed]

39. Yao, Y.G.; Lin, Z.Y.; Li, Z.; Song, X.J.; Moona, K.S.; Wong, C.P. Large-scale production of two-dimensional nanosheets. *J. Mater. Chem.* **2012**, *22*, 13494–13499. [CrossRef]

40. Gopalakrishnan, D.; Damien, D.; Shaijumon, M.M. MoS$_2$ quantum dot-interspersed exfoliated MoS$_2$ Nanosheets. *ACS Nano* **2014**, *8*, 5297–5303. [CrossRef] [PubMed]

41. Shen, J.; He, Y.; Wu, J.; Gao, C.; Keyshar, K.; Zhang, X.; Yang, Y.; Ye, M.; Vajtai, R.; Lou, J.; et al. Liquid phase exfoliation of two-dimensional materials by directly probing and matching surface tension components. *Nano Lett.* **2015**, *15*, 5449–5454. [CrossRef] [PubMed]

42. Nicolosi, V.; Chhowalla, M.; Kanatzidis, M.G.; Strano, M.S.; Coleman, J.N. Liquid exfoliation of layered materials. *Science* **2013**, *340*, 1421–1439. [CrossRef]

43. Splendiani, A.; Sun, L.; Zhang, Y.B.; Li, T.S.; Kim, J.; Chim, C.Y.; Galli, G.; Wang, F. Emerging photoluminescence in Monolayer MoS$_2$. *Nano Lett.* **2010**, *10*, 1271–1275. [CrossRef] [PubMed]

44. Chen, Z.; Forman, A.J.; Jaramillo, T.F. Bridging the gap between bulk and nanostructured photoelectrodes: the impact of surface states on the electrocatalytic and photoelectrochemical properties of MoS$_2$. *J. Phys. Chem.C* **2013**, *117*, 9713–9722. [CrossRef]

 *nanomaterials*

Article

Operation Mechanism of a MoS_2/BP Heterojunction FET

Sung Kwan Lim [1,2], Soo Cheol Kang [1,3], Tae Jin Yoo [1,3], Sang Kyung Lee [1,3], Hyeon Jun Hwang [1,3] and Byoung Hun Lee [1,3,*]

[1] Center for Emerging Electronic Devices and Systems (CEEDS), GIST, 123 Cheomdan-gwagiro, Buk-gu, Gwangju 61005, Korea; lsk8410@gist.ac.kr (S.K.L.); soocheol@gist.ac.kr (S.C.K.); tjyoo123@gist.ac.kr (T.J.Y.); leesk@gist.ac.kr (S.K.L.); hhjune@gist.ac.kr (H.J.H.)

[2] Department of Nanobio Materials and Electronics, GIST, 123 Cheomdan-gwagiro, Buk-gu, Gwangju 61005, Korea

[3] School of Materials and Science Engineering, GIST, 123 Cheomdan-gwagiro, Buk-gu, Gwangju 61005, Korea

*   Correspondence: bhl@gist.ac.kr; Tel.: +82-(0)62-715-2347

Received: 20 September 2018; Accepted: 4 October 2018; Published: 7 October 2018

Abstract: The electrical characteristics and operation mechanism of a molybdenum disulfide/black phosphorus (MoS_2/BP) heterojunction device are investigated herein. Even though this device showed a high on-off ratio of over 1×10^7, with a lower subthreshold swing of ~54 mV/dec and a 1fA level off current, its operating mechanism is closer to a junction field-effect transistor (FET) than a tunneling FET. The off-current of this device is governed by the depletion region in the BP layer, and the band-to-band tunneling current does not contribute to the rapid turn-on and extremely low off-current.

Keywords: MoS_2; black phosphorus; 2D/2D heterojunction; junction FET; tunneling diode; tunneling FET; band-to-band tunneling (BTBT)

---

1. Introduction

Tunneling field-effect transistors (tFETs) have been studied as an alternative device for silicon MOSFET enabling very sharp turn-on which is required to reduce the operation voltage and the system power consumption. tFETs utilize band-to-band tunneling (BTBT) from a source to a channel, and an off-current is maintained using a P-N-N or N-P-P-type channel-doping profile [1–5]. When BTBT occurs in this channel-doping profile, the carriers from the source are injected directly into the channel and transported to the drain. When BTBT is not possible, the carrier cannot be injected into the drain because of the barrier formed in the channel region. In this device, the tunneling distance should be minimized to allow the tunneling current to rapidly increase. Thus, a very sharp P-N junction should be formed. The performances of experimental tunnel FETs reported in the literature have not reached their theoretical limit, primarily due to graded doping profiles and interface traps [3,6]. For an ideal BTBT current flow, an atomically sharp interface with minimal interface states is necessary. Fortunately, these requirements can be easily satisfied using transition metal dichalcogenide (TMD) materials because the various choices of band gaps and band alignment combinations make the stack of two-dimensional (2D) materials an ideal candidate for tunneling FETs [7–11]. Thus, a variety of stacks, including molybdenum disulfide (MoS_2)/tungsten diselenide (WSe_2), tin diselenide ($SnSe_2$)/WSe_2, MoS_2/black phosphorus (BP), and $SnSe_2$/BP have been investigated [12–19]. Most of these studies explain that the turn-on mechanism is due to the BTBT, and the turn off mechanism is due to the band misalignment.

In this work, we fabricated a heterojunction FET, using a multilayer MoS_2 and a thick black-phosphorus stack with a back gate structure, and investigated the operation mechanism.

This system was chosen because a MoS_2/BP stack is suitable for broken bandgap device fabrication. Our analysis revealed that the operation mechanism of this heterojunction FET is quite different from what has been reported in the literature. The off-current is dominated by the depletion in the BP layer, and the subthreshold swing is related to the reduction of the depletion region. The BTBT current only contributes to the hump in the drain current.

2. Materials and Methods

The fabrication processes of the MoS_2/BP heterojunction FET are shown in Figure 1. Figure 1a shows the structure of a stamp used to transfer 2D flakes. Polypropylene carbonate (PPC) (Sigma-Aldrich, CAS 25511-85-7, Sigma-Aldrich, CAS 25511-85-7, St. Louis, MO, USA) was used to pick up and transfer the flakes of MoS_2 and BP at a low temperature [18,20,21]. Since the flakes are easily damaged during the detachment process, and some of 2D materials, e.g., $SnSe_2$, hafnium diselenide ($HfSe_2$), and BP, can be oxidized during the transfer or device fabrication [22–24], we modified the fabrication process to directly transfer the flakes to the PPC film to minimize the damage and to reduce the air exposure time. Both sides of a handmade polydimethylsiloxane (PDMS) sheet were treated with ozone plasma for 10 min to improve the adhesion of the double-sided tape to the PDMS sheet. The PPC film (15% solution in Anisole) was coated onto the stack of tape/PDMS/tape and cured on a hot plate at 100 °C for 10 min. Then, the PPC/tape/PDMS/tape sheet was placed on a glass slide patterned with align keys. Figure 1b–d show the rest of the device fabrication process.

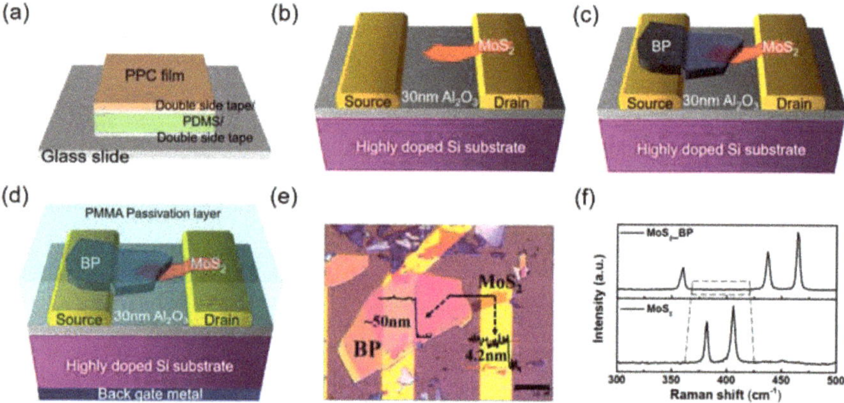

Figure 1. (**a**) Schematic of stamp (polypropylene carbonate (PPC)/double-sided tape/ polydimethylsiloxane (PDMS)/double-sided tape/glass slide), with the 2D flake transferred directly onto the PPC film. (**b**) MoS_2 transferred onto the drain electrode (5-nm/45-nm Ti/Au) and gate oxide (30-nm Al_2O_3). (**c**) Black phosphorus (BP) flake transferred quickly to the substrate using the same method. (**d**) Device passivated using polymethylmethacrylate (PMMA) film. (**e**) Optical image of MoS_2/BP (4.2 nm/50 nm) heterojunction. (**f**) The thickness of flakes was measured using Raman spectra (using a 514-nm laser) of the molybdenum disulfide (MoS_2)/BP stack. The lower panel shows the Raman spectra of the MoS_2 flake (the E_{2g}^1 peak at 382.29 cm^{-1} and the A_{1g} peak at 406.25 cm^{-1}).

The source and drain electrodes (5-nm Ti/45-nm Au) were formed on a 30-nm aluminum oxide (Al_2O_3)/highly doped P-type silicon substrate using e-beam evaporation and photolithography. In this experiment, MoS_2 was used as the channel material with BP as the source material. Exfoliated MoS_2 flakes were transferred to the PPC film from a bulk crystal using commercial adhesive tape, and then transferred onto the drain electrodes using a dry transfer system at 80 °C. The selected BP flake was also transferred onto the source electrode using the same process, while carefully overlapping the BP flake onto the MoS_2 flake that was already connected to the drain electrode. Since the BP flake could be easily oxidized in air [24], a polymethylmethacrylate (PMMA, 950 K 4 A, Microchem, Westborough,

MA, USA) coating was applied, followed by thermal annealing at 180 °C for 5 min to eliminate the solvent. Figure 1e shows an optical microscope image of the device. The thickness of the MoS_2, measured with atomic force microscopy, was 4.2 nm and the BP thickness was ~50 nm. Figure 1f shows the Raman spectrum of the BP/MoS_2, measured from the overlapped region. The characteristic BP peaks were observed at 360.65 (A_g^1), 437.3 (B_{2g}), and 464.4 (A_g^2) cm^{-1}; however, the Raman peak of the MoS_2 was not observed in this spectrum because the BP layer was very thick. The Raman spectrum of the MoS_2 shown in the lower part of Figure 1f was measured from the region not overlapping the BP layer.

3. Results and Discussion

First, the electrical characteristics of the MoS_2/BP diode were measured using a parameter analyzer (Keithely 4200, Santa Rosa, CA, USA). The TMD materials show different electrical characteristics depending on the thickness of the layer. When the BP is very thick, no significant gate modulation is observed, as shown in Figure 2a. In fact, this characteristic is beneficial for device operation because the high current level with a small gate modulation means that the BP layer can be used as a good contact material with a bandgap.

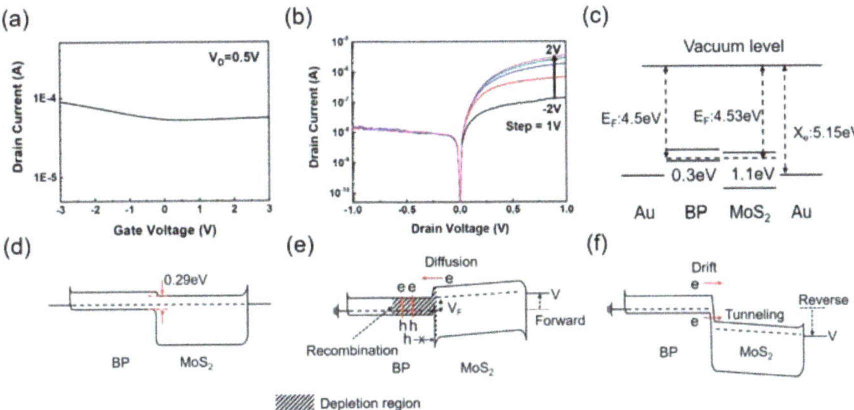

Figure 2. (**a**) Transfer characteristic of the thick-layer BP field-effect transistor (FET). (**b**) Electrical characteristics of a MoS_2/BP diode following the gate voltage. Band structure of the MoS_2/BP, (**c**) before contact, (**d**) at the equilibrium state, and (**e**) with a forward rectifying condition with negative bias applied to the MoS_2 electrode. The holes from BP cannot overcome the high barrier at the forward bias and the electrons from MoS_2 diffuses into BP, generating a depletion region. (**f**) Reverse bias condition with positive bias applied to the MoS_2 electrode. The current is primarily due to the drift of minority carriers, as well as the tunneling carriers from the BP side.

Figure 2b shows the diode characteristics of the MoS_2/BP heterojunction at different gate biases, from −2 to 2 V with a gate bias step of 1 V. While the potential of the BP layer is almost fixed to the source–drain bias, the Fermi level of the MoS_2 layer shows a reasonable gate modulation for both single layers and multilayers [25]. As the gate bias increased from −2 V, the rectification characteristics at the MoS_2/BP junction seemed to improve because the barrier height at the MoS_2/BP interface increased. These characteristics can be explained more intuitively with a band diagram. The ideal band structure of a MoS_2/BP stack before stacking is shown in Figure 2c. The work function of the 2D materials is measured differently depending on the measurement environment due to its high surface energy. We assumed that the Fermi levels of the MoS_2 and BP are 4.53 and 4.5 eV, respectively [26,27]. After the stacking, the MoS_2/BP heterojunction forms a staggered (type II) band alignment at an equilibrium state, with a very small barrier on the conduction band side, as shown in Figure 2d. Theoretically, the effective band gap, which is the difference between the conduction band of MoS_2 and the valence band of BP, formed at the MoS_2/BP junction is 0.29 eV. The effective band gap is modulated by the drain bias during the diode type operation.

Even though the doping profile of a MoS_2/BP junction is similar to a P-N junction, the carrier conduction mechanisms are quite different. When a negative drain bias is applied, the Fermi level of the MoS_2 shifts upward (forward bias for a P-N junction) and the majority carriers from MoS_2 flow into the BP layer; however, the holes in the BP layer cannot be transferred to the MoS_2 layer, due to the high barrier height. As a result, the recombination of electrons and holes at the BP side generates a depletion region, which is balanced by the electron influx and the resistance increase, due to the depletion width increase. Hence, an almost constant current of ~10 nA is maintained in our device. In the case of a silicon P-N junction, the current increases exponentially at forward bias.

On the other hand, when the drain bias is positive (reverse bias for a P-N junction), minority carriers from the BP and MoS_2 layers start to flow to opposite sides, driven by the electric field. Moreover, depending on the drain bias, the tunneling component may also contribute to the drain current. The current flow, shown in Figure 2b, saturates at a high drain bias because the current flow is limited by the minority carrier supply. Unlike a P-N junction, where the diffusion of the majority carriers is the primary conduction mechanism, the drift of minority carriers is the primary conduction mechanism in this bias region. Many prior studies have correctly noted this difference; however, in our opinion, they did not carefully consider the off-current mechanism [13,14,16,18,19]. Most of prior works explained that the off state is due to the band shift closing the direct tunneling window, but they did not consider that the gate bias region—causing extremely low off current—did not match the gate bias region of the direct tunneling current.

Figure 3a shows the transfer characteristics of a MoS_2/BP FET with a small positive drain bias. In this case, the current level is already in the 10 to 100 nA range at V_G = 0 V and V_D = 500 mV, as shown in Figures 2b and 3a. Thus, to turn off this device, a strong negative gate bias should be applied. Many prior studies described band structures similar to Figure 2d to explain the off-state, where the tunneling current does not flow because the carriers in the BP valence band cannot be transferred to the MoS_2 conduction band. However, as indicated in Figure 2c, the minority carriers from the BP can be injected into the MoS_2 (and vice versa) when V_G is approximately −3 V and the current level is approximately 10 nA. Thus, the reduction of the tunneling component cannot explain the turn off mechanism of our device; i.e., the prior explanation is obviously wrong.

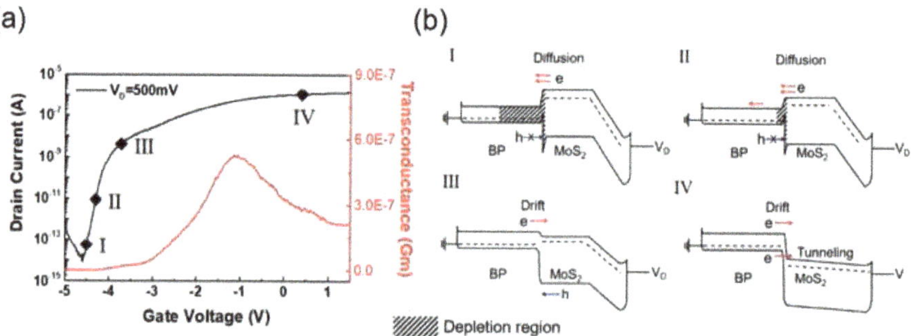

Figure 3. Electrical properties and current flow mechanism of a MoS_2/BP heterojunction FET at V_D = 500 mV. (**a**) Transfer characteristics and transconductance (G_m). (**b**) Band diagrams showing the states at different gate bias regions.

Thus, the extremely low off current at strong negative gate bias needs to be explained with another mechanism. If we think about the carrier conduction at very negative V_G, only the electrons from MoS_2 can be drifted into the BP region and hole drift is blocked by the high barrier as shown in Figure 3b. Then, electrons injected into the BP region recombines holes and form a depletion region. As the

V_G becomes more negative, the width of depletion region increases further and the drain current decreases rapidly, until the hole diffusion current starts to increase at $V_G < -4.5$ V. Thus, in our opinion, the off-current of the MoS$_2$/BP FET can be better explained with the formation of a depletion region in the BP layer.

If we explain the device operation from the negative V_G side, it is easier to understand the operating mechanism. The drain current does not flow at -4.5 V because of the large depletion width. As V_G increases to the positive bias side, the depletion region decreases, and suddenly, the minority carriers start drifting to other materials. Then, as the MoS$_2$ energy band moves further downward, the tunneling current starts to flow at -3.2 V. Since the tunneling current is added to the drift current, due to the minority carrier injection, the drain current shows a hump at -3.2 V in our device. Most tunnel FETs reported in the literature show this kind of hump in the transfer characteristics, confirming our model.

The transfer characteristics measured at different drain biases and temperatures support our operation mechanism model further. When V_D is small, the drift current decreases, but the turn-on behavior is not strongly affected because it is more closely related to how the depletion region is formed by the initial band alignment at the BP and MoS$_2$ interface. To detect the bias where the BTBT current starts to contribute, the second derivative of the transfer curves is calculated, as shown in Figure 4a. The starting point of the abrupt curvature change marked with an arrow indicates the point of the BTBT current initiation, and the peak position indicates the maximum tunneling current. As the drain bias increases, the band alignment approaches the state shown in Figure 2f. Thus, a higher negative V_G should be applied to turn off the tunneling current by pushing the MoS$_2$ energy band upward. The temperature dependence also shows an interesting characteristic. The position of tunneling current initiation has not changed significantly, but the peak height increased, indicating that the BTBT current increased due to the increase of thermally activated carriers in the valence band of BP. The drift current increase can be attributed to the increased minority carrier density at the higher temperature.

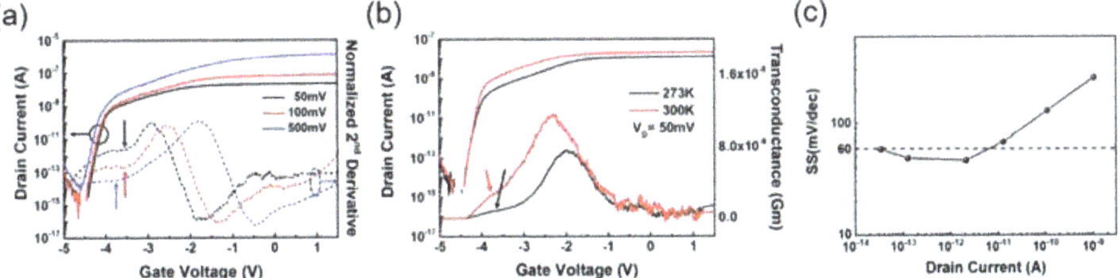

Figure 4. (**a**) Transfer characteristics of a MoS$_2$/BP heterojunction FET for different drain voltages (50 mV, 100 mV, and 500 mV). Normalized second derivative of transfer curves are shown to note the initiation points of band-to-band tunneling (BTBT). (**b**) Temperature-dependent transfer characteristic at 273 and 300 K, V_D = 50 mV. (**c**) Subthreshold swing (SS) versus drain current at V_D = 500 mV, 300 K.

Finally, we would like to emphasize that our device also shows a sub-60 mV/dec subthreshold swing in some regions of the transfer curve, as shown in Figure 4c. In previous reports (Table 1), swing values below 60 mV/dec often suggest a tunneling mechanism [12,13,16–19]. However, 60 mV/dec is the limit set by the diffusion mechanism. Since we have proposed that the turn-on behavior of our device is governed by the formation of the depletion region at the MoS$_2$/BP interface, the turn-on mechanism is closer to a junction FET, where the drain current starts to flow once a small current path is formed by the reduction of the depletion width. Thus, the swing that is smaller than 60 mV/dec is more closely related to geometric factors and the carrier profile at the BP region, which affect the shape of depletion region.

Table 1. Comparison of the performance of the 2D/2D tunneling FETs reported in the literature.

Ref.	Material	On Current (A) (V_D)	I_{on}/I_{off} Ratio	SS_{MIN} (mV/dec) at RT	SS_{AVG} (mV/dec) at RT	Dielectric
Our result	MoS_2/BP	1×10^{-6} (500 mV)	~7×10^7	54	94	30-nm Al_2O_3 (bottom)
[14]	MoS_2/p-Ge	5×10^{-6} (500 mV)	~8×10^7	3.9	22	Ion gel (top)
[16]	MoS_2/BP	8×10^{-6} (50 mV)	10^6	55	55	Ion gel (top)
[18]	MoS_2/WSe_2	-	-	-	75	10-nm HfO_2 (bottom)
[19]	MoS_2/BP	1×10^{-7} (800 mV)	~10^4	-	65	Ion gel (top)
[22]	WSe_2/$SnSe_2$	9×10^{-7} (500 mV)	~10^5	37	80	40-nm Al_2O_3 (bottom)

4. Conclusions

In conclusion, we demonstrated the MoS_2/BP heterojunction FET and analyzed the device operation mechanism. We found that the BTBT is not the primary mechanism determining the on-off characteristics of the MoS_2/BP heterojunction FET, but it contributes to the formation of the hump in the transfer curve. In addition, the rapid turn-on and extremely low off-current are explained by the depletion region formation. Our results can be applied to general 2D/2D heterojunction devices.

Author Contributions: S.K.L. designed and conducted the experiments and S.C.K. supported the electrical measurement and analysis. T.J.Y. set up the experiment system. S.K.L. and H.J.H supported the process of experiments and the analysis of data. B.H.L. supported and guided the experiment and the results. B.H.L. conceived and advised the publication of the paper.

Funding: This work was partially supported by the Nano Materials Technology Development Program (2016M3A7B4909942) and by the Creative Materials Discovery Program of the Creative Multilevel Research Center (2015M3D1A1068062) through the National Research Foundation (NRF) of Korea funded by the Ministry of Science and ICT.

Conflicts of Interest: The authors declare no conflict of interest.

References

1. Ionescu, A.M.; Riel, H. Tunnel field-effect transistors as energy-efficient electronic switches. *Nature* **2011**, *479*, 329–337. [CrossRef] [PubMed]

2. Choi, W.Y.; Park, B.G.; Lee, J.D.; Liu, T.J.K. Tunneling Field-Effect Transistors (TFETs) With Subthreshold Swing (SS) Less Than 60 mV/dec. *IEEE Electron. Device Lett.* **2007**, *28*, 743–745. [CrossRef]

3. Morita, Y.; Mori, T.; Migita, S.; Mizubayashi, W.; Tanabe, A.; Fukuda, K.; Matsukawa, T.; Endo, K.; O'uchi, S.; Liu, Y.; et al. Synthetic electric field tunnel FETs: Drain current multiplication demonstrated by wrapped gate electrode around ultrathin epitaxial channel. In Proceedings of the 2013 Symposium on VLSI Technology, Kyoto, Japan, 11–13 June 2013; pp. T236–T237.

4. Knoll, L.; Zhao, Q.; Nichau, A.; Trellenkamp, S.; Richter, S.; Schäfer, A.; Esseni, D.; Selmi, L.; Bourdelle, K.K.; Mantl, S. Inverters With Strained Si Nanowire Complementary Tunnel Field-Effect Transistors. *IEEE Electron. Device Lett.* **2013**, *34*, 813–815. [CrossRef]

5. Kim, M.; Wakabayashi, Y.; Nakane, R.; Yokoyama, M.; Takenaka, M.; Takagi, S. High I_{on}/I_{off} Ge-source ultrathin body strained-SOI tunnel FETs. In Proceedings of the 2014 IEEE International Electron Devices Meeting, San Francisco, CA, USA, 5–17 December 2014; pp. 13.2.1–13.2.4.

6. Vandooren, A.; Leonelli, D.; Rooyackers, R.; Hikavyy, A.; Devriendt, K.; Demand, M.; Loo, R.; Groeseneken, G.; Huyghebaert, C. Analysis of trap-assisted tunneling in vertical Si homo-junction and SiGe hetero-junction Tunnel-FETs. *Solid State Electron.* **2013**, *83*, 50–55. [CrossRef]

7. Radisavljevic, B.; Radenovic, A.; Brivio, J.; Giacometti, V.; Kis, A. Single-layer MoS_2 transistors. *Nat. Nanotechnol.* **2011**, *6*, 147–150. [CrossRef] [PubMed]

8. Lee, C.-H.; Lee, G.-H.; van der Zande, A.M.; Chen, W.; Li, Y.; Han, M.; Cui, X.; Arefe, G.; Nuckolls, C.; Heinz, T.F. Atomically thin p–n junctions with van der Waals heterointerfaces. *Nat. Nanotechnol.* **2014**, *9*, 676–681. [CrossRef] [PubMed]

9. Chuang, H.-J.; Tan, X.; Ghimire, N.J.; Perera, M.M.; Chamlagain, B.; Cheng, M.M.-C.; Yan, J.; Mandrus, D.; Tománek, D.; Zhou, Z. High mobility WSe$_2$ p- and n-type field-fffect fransistors contacted by highly doped graphene for low-resistance contacts. *Nano Lett.* **2014**, *14*, 3594–3601. [CrossRef] [PubMed]

10. Das, S.; Appenzeller, J. WSe$_2$ field effect transistors with enhanced ambipolar characteristics. *Appl. Phys. Lett.* **2013**, *103*, 103501. [CrossRef]

11. Yu, W.J.; Li, Z.; Zhou, H.; Chen, Y.; Wang, Y.; Huang, Y.; Duan, X. Vertically stacked multi-heterostructures of layered materials for logic transistors and complementary inverters. *Nat. Mater.* **2012**, *12*, 246–252. [CrossRef] [PubMed]

12. Roy, T.; Tosun, M.; Cao, X.; Fang, H.; Lien, D.-H.; Zhao, P.; Chen, Y.-Z.; Chueh, Y.-L.; Guo, J.; Javey, A. Dual-Gated MoS$_2$/WSe$_2$ van der Waals Tunnel Diodes and Transistors. *ACS Nano* **2015**, *9*, 2071–2079. [CrossRef] [PubMed]

13. Roy, T.; Tosun, M.; Hettick, M.; Ahn, G.H.; Hu, C.; Javey, A. 2D-2D tunneling field-effect transistors using WSe$_2$/SnSe$_2$ heterostructures. *Appl. Phys. Lett.* **2016**, *108*, 083111. [CrossRef]

14. Sarkar, D.; Xie, X.; Liu, W.; Cao, W.; Kang, J.; Gong, Y.; Kraemer, S.; Ajayan, P.M.; Banerjee, K. A subthermionic tunnel field-effect transistor with an atomically thin channel. *Nature* **2015**, *526*, 91–95. [CrossRef] [PubMed]

15. Yan, R.; Fathipour, S.; Han, Y.; Song, B.; Xiao, S.; Li, M.; Ma, N.; Protasenko, V.; Muller, D.A.; Jena, D.; et al. Esaki Diodes in van der Waals Heterojunctions with Broken-Gap Energy Band Alignment. *Nano Lett.* **2015**, *15*, 5791–5798. [CrossRef] [PubMed]

16. Liu, X.; Qu, D.; Li, H.-M.; Moon, I.; Ahmed, F.; Kim, C.; Lee, M.; Choi, Y.; Cho, J.H.; Hone, J.C.; et al. Modulation of Quantum Tunneling via a Vertical Two-Dimensional Black Phosphorus and Molybdenum Disulfide p–n Junction. *ACS Nano* **2017**, *11*, 9143–9150. [CrossRef] [PubMed]

17. Shim, J.; Oh, S.; Kang, D.-H.; Jo, S.-H.; Ali, M.H.; Choi, W.-Y.; Heo, K.; Jeon, J.; Lee, S.; Kim, M.; et al. Phosphorene/rhenium disulfide heterojunction-based negative differential resistance device for multi-valued logic. *Nat. Commun.* **2016**, *7*, 13413. [CrossRef] [PubMed]

18. Nourbakhsh, A.; Zubair, A.; Dresselhaus, M.S.; Palacios, T. Transport Properties of a MoS$_2$/WSe$_2$ Heterojunction Transistor and Its Potential for Application. *Nano Lett.* **2016**, *16*, 1359–1366. [CrossRef] [PubMed]

19. Xu, J.; Jia, J.; Lai, S.; Ju, J.; Lee, S. Tunneling field effect transistor integrated with black phosphorus-MoS$_2$ junction and ion gel dielectric. *Appl. Phys. Lett.* **2017**, *110*, 033103. [CrossRef]

20. Pizzocchero, F.; Gammelgaard, L.; Jessen, B.S.; Caridad, J.M.; Wang, L.; Hone, J.; Bøggild, P.; Booth, T.J. The hot pick-up technique for batch assembly of van der Waals heterostructures. *Nat. Commun.* **2016**, *7*, 11894. [CrossRef] [PubMed]

21. Wang, L.; Meric, I.; Huang, P.Y.; Gao, Q.; Gao, Y.; Tran, H.; Taniguchi, T.; Watanabe, K.; Campos, L.M.; Muller, D.A.; et al. One-Dimensional Electrical Contact to a Two-Dimensional Material. *Science* **2013**, *342*, 614–617. [CrossRef] [PubMed]

22. Yan, X.; Liu, C.; Li, C.; Bao, W.; Ding, S.; Zhang, D.W.; Zhou, P. Tunable SnSe$_2$/WSe$_2$ Heterostructure Tunneling Field Effect Transistor. *Small* **2017**, *13*, 1701478. [CrossRef] [PubMed]

23. Kang, M.; Rathi, S.; Lee, I.; Lim, D.; Wang, J.; Li, L.; Khan, M.A.; Kim, G.-H. Electrical characterization of multilayer HfSe$_2$ field-effect transistors on SiO$_2$ substrate. *Appl. Phys. Lett.* **2015**, *106*, 143108. [CrossRef]

24. Wood, J.D.; Wells, S.A.; Jariwala, D.; Chen, K.-S.; Cho, E.; Sangwan, V.K.; Liu, X.; Lauhon, L.J.; Marks, T.J.; Hersam, M.C. Effective Passivation of Exfoliated Black Phosphorus Transistors against Ambient Degradation. *Nano Lett.* **2014**, *14*, 6964–6970. [CrossRef] [PubMed]

25. Chu, L.; Schmidt, H.; Pu, J.; Wang, S.; Özyilmaz, B.; Takenobu, T.; Eda, G. Charge transport in ion-gated mono-, bi-, and trilayer MoS$_2$ field effect transistors. *Sci. Rep.* **2014**, *4*, 7293. [CrossRef] [PubMed]

26. Shakya, J.; Kumar, S.; Kanjilal, D.; Mohanty, T. Work Function Modulation of Molybdenum Disulfide Nanosheets by Introducing Systematic Lattice Strain. *Sci. Rep.* **2017**, *7*, 9576. [CrossRef] [PubMed]

27. Cai, Y.; Zhang, G.; Zhang, Y.-W. Layer-dependent Band Alignment and Work Function of Few-Layer Phosphorene. *Sci. Rep.* **2015**, *4*, 6677. [CrossRef] [PubMed]

 *nanomaterials*

Communication

Exfoliation and Characterization of V_2Se_9 Atomic Crystals

Bum Jun Kim [1,†], **Byung Joo Jeong** [2,†], **Seungbae OH** [2], **Sudong Chae** [2], **Kyung Hwan Choi** [1], **Tuqeer Nasir** [1], **Sang Hoon Lee** [2], **Kwan-Woo Kim** [2], **Hyung Kyu Lim** [2], **Ik Jun Choi** [2], **Ji-Yun Moon** [3], **Hak Ki Yu** [3], **Jae-Hyun Lee** [3,*] and **Jae-Young Choi** [1,2,*]

1   SKKU Advanced Institute of Nanotechnology, Sungkyunkwan University, Suwon 16419, Korea; kbj454@skku.edu (B.J.K.); chhcc12@gmail.com (K.H.C.); tuqeernasir166@gmail.com (T.N.)
2   School of Advanced Materials Science and Engineering, Sungkyunkwan University, Suwon 16419, Korea; jbj929@skku.edu (B.J.J.); nysbo0219@gmail.com (S.O.); csd5432@gmail.com (S.C.); alfhdj@gmail.com (S.H.L.); rhksdn8904@gmail.com (K.-W.K.); hyungkyu1992@gmail.com (H.K.L.); cksoon16@gmail.com (I.J.C.)
3   Department of Materials Science and Engineering and Department of Energy Systems Research, Ajou University, Suwon 16499, Korea; ydnas96@ajou.ac.kr (J.-Y.M.); hakkiyu@ajou.ac.kr (H.K.Y.)
*   Correspondence: jaehyunlee@ajou.ac.kr (J.-H.L.); jy.choi@skku.edu (J.-Y.C.); Tel.: +82-(0)31-219-2465 (J.-H.L.); +82-(0)31-290-7353 (J.-Y.C.)
†   These authors contributed equally to this work.

Received: 31 August 2018; Accepted: 18 September 2018; Published: 18 September 2018

Abstract: Mass production of one-dimensional, V_2Se_9 crystals, was successfully synthesized using the solid-state reaction of vanadium and selenium. Through the mechanical exfoliation method, the bulk V_2Se_9 crystal was easily separated to nanoribbon structure and we have confirmed that as-grown V_2Se_9 crystals consist of innumerable single V_2Se_9 chains linked by van der Waals interaction. The exfoliated V_2Se_9 flakes can be controlled thickness by the repeated-peeling method. In addition, atomic thick nanoribbon structure of V_2Se_9 was also obtained on a 300 nm SiO_2/Si substrate. Scanning Kelvin probe microscopy analysis was used to explore the variation of work function depending on the thickness of V_2Se_9 flakes. We believe that these observations will be of great help in selecting suitable metal contacts for V_2Se_9 and that a V_2Se_9 crystal is expected to have an important role in future nano-electronic devices.

Keywords: V_2Se_9; atomic crystal; mechanical exfoliation; scanning Kelvin probe microscopy

---

1. Introduction

To overcome the high-density integration of electronic technology, which faces physical limitations (e.g., fabrication process and reduction in charge carrier mobility), researchers have been intensively trying to develop a new device architecture or novel materials [1–4]. A range of diverse candidate materials have been proposed since the 2000s. Among them, graphene, which is a single layer of carbon atoms arranged in a hexagonal lattice, is considered to be a promising solution for future electronic devices because of its superior physical properties such as high carrier mobility and excellent chemical stability; however, it has the fatal disadvantage in that it has difficulty forming a band gap [1,2,5–7]. Therefore, the development of applications for graphene-based electronic devices, the most promising field, does not meet public' expectation yet. Graphene nanoribbons (GNRs) are presented as the effective way to open the bandgap of graphene but it is difficult to produce a uniform width in large area [6,7]. In addition, the transport behavior of GNRs and newly introduced two-dimensional (2D) materials (e.g., transition metal dichalcogenides (TMDCs) and black phosphorous), with appropriate bandgaps, are reduced dramatically because of dangling bonds at the side edges and domain boundaries [8–11]. Unfortunately, most of the studies of the

2D material-based electronic devices thus far contain an etching process to define the conducting channel. Thus, the discovery of one-dimensional (1D) nanomaterials, which are free from edge and grain boundary scattering, is a key solution in the development of nano-electronic device.

Carbon nanotubes (CNTs), which exhibit high carrier mobility, ultimate mechanical strength, and chemical stability, have been considered as representative building blocks for next-generation transistors, chemical sensors, and nanocomposites [12–14]. However, the wide range of electronic structures that arise from the different chirality of the CNTs curtails the reliability of the manufacturing process of the nano-electronic devices [15]. Therefore, separation of single-chirality CNTs from the bulk CNTs or control of the chirality during the growth of the CNTs is required. Recently, studies on the synthesis and application of a new family of 1D nanomaterials in the form of three-dimensional (3D) bundles of numerous single-molecular chains coupled by weak van der Waals interactions have been reported [16–21]. For example, extensive studies on bulk synthesis and atomic-scale dispersion of the bio-compatible $Mo_6S_{9-x}I_x$ have been reported [21–23]. In addition, Sb_2S_3 was developed as an optoelectronic device by effectively reducing exciton decay due to the absence of dangling bonds [24]. Moreover, VS_4 was utilized for an electrochemical energy storage device by using the van der Waals gap between the chains [25,26]. However, in the majority of studies on these materials, they have been utilized only as a thin-film structure, although the benefits of the layered characteristics can be exploited. In addition, the crystal structure of $Mo_6S_{9-x}I_x$ is not well defined because the position of the sulfur and iodine atoms bridged to the molybdenum atoms may vary even for the same stoichiometric composition.

In this study, we succeeded in mass producing 1D semiconductor V_2Se_9 crystals via a simple transport method. Through the mechanical exfoliation method, we confirmed that as-grown V_2Se_9 crystals consist of innumerable single V_2Se_9 chains linked via the van der Waals interaction, like graphite. In addition, a nanoribbons structure of V_2Se_9 which is capable of thickness control was obtained through repetitive mechanical exfoliation of the V_2Se_9 crystals. Lastly, the change in work function according to the thickness change of the V_2Se_9 flakes was analyzed by scanning Kelvin probe microscopy (SKPM) measurement.

2. Materials and Methods

Synthesis: V_2Se_9 was synthesized using V (Powder, −325 mesh, 99.5%, Sigma-Aldrich, St. Louis, MO, USA) and Se (powder, 99+%, Alfa Aesar, Haverhill, MA, USA). The mixture of V (0.2038 g) and Se (1.4213 or 1.9898 g) with a V to Se ratio of 2:9 or 2:12.6 was pelletized and then sealed in a 10 cm-long evacuated quartz tube. The quartz ampoule was heated for 120 h at a temperature of 300–400 °C (at 5.5 °C/h) and then cooled (at 10 °C/h). The resulting material was a dark gray sintered powder. The unreacted Se was sublimated by heat treatment in a tube furnace at 250 °C under Ar atmosphere for 24 h.

Mechanical exfoliation: The bulk V_2Se_9 was placed on wafer dicing tape (BT150EKL, Nitto Denko, Umeda, Osaka, Japan) and the materials were stuck several times to yield thinner-than-bulk materials. A substrate (300 nm SiO_2/Si or bare Si) was cleaned by ultrasonication in acetone, ethanol, and DI water for 15 min, followed by heating at 100 °C in order to remove the moisture from the substrate. The polymer tape was adhered strongly to and pressed against the substrate. After adhesion, the polymer tape was removed from the substrate; this process was repeated for exfoliation.

Characterization: Powder X-ray diffraction (Mac Science, M18XHF22, Tokyo, Japan) was performed using Cu-Kα radiation (λ = 0.154 nm). Field emission-scanning electron microscopy (FE-SEM, Hitachi, S4300SE, Chiyoda, Tokyo, Japan) was operated at an acceleration voltage of 15 kV. Atomic force microscopy (AFM, Park systems, NX 10, Suwon, South Korea) was performed in a non-contact mode for the topographic analysis of the mechanically exfoliated V_2Se_9 on 300 nm Si/SiO_2. The surface potentials of V_2Se_9 on Si substrate were measured by SKPM (Park systems, NX10, Suwon,

South Korea) measurement using Si tips coated with Cr-Pt (Multi75-G, Budget Sensors Inc., 1113 Sofia, Bulgaria) with resonance frequencies of 75 kHz, a scan rate of 0.3 Hz, and sample bias of ± 1 V.

3. Results and Discussion

Since the transition metal vanadium has the outermost 3d orbital, it can produce various forms of compounds (e.g., V_5Se_4 to V_2Se_9) through a chemical reaction with selenium (see the phase diagram in Figure S1). Therefore, to synthesize V_2Se_9 crystals with a high-purity and high-crystallinity, the ratio of V:Se and the synthesis temperature should be considered carefully. For example, if the atomic mixing ratio of V and Se powder is adjusted precisely to 2:9 to synthesize V_2Se_9 crystals, unpredictable fluctuation occurs in the synthetic tube and VSe_2, which is an undesirable impurity, is formed. We corrected these parameters experimentally, and as a result obtained pure V_2Se_9 crystals with an exact stoichiometry ratio of 2:9 by adding them in excess of Se, as shown in Figure 1a (V:Se atomic mixing ratio of 2:12.6). The crystallinity of the bulk V_2Se_9 crystal was verified by the X-ray diffraction (XRD) pattern (JCPDS 01-077-1589) (Figure 1b). The SEM images in Figure 1c,d clearly shows the 1D nanowire structures and the gaps generated during transfer of the sample onto the Si substrate.

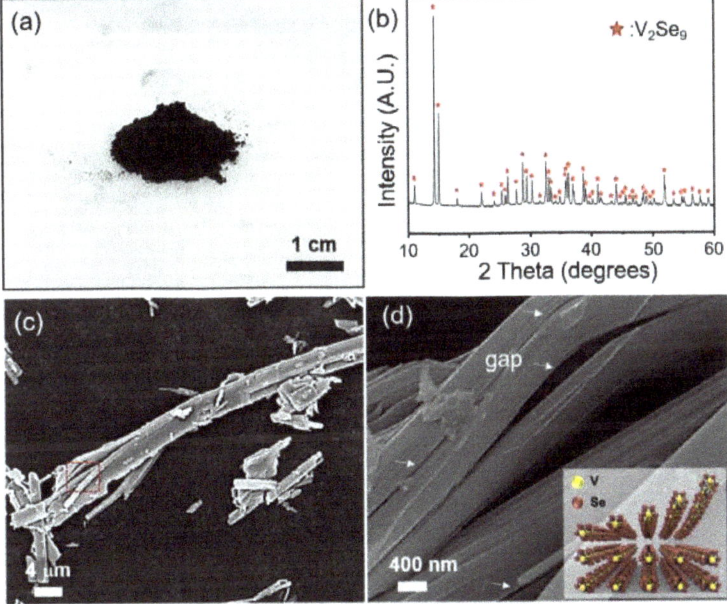

Figure 1. (**a**) Photo-image of mass production of V_2Se_9 crystal. (**b**) XRD pattern of V_2Se_9 crystal. (**c**) Low- and (**d**) high-magnification SEM images of V_2Se_9 crystal. The inset shows an illustration of the crystal structure of V_2Se_9.

To investigate the structural characteristics of nanoscale V_2Se_9, the bulk V_2Se_9 crystal was mechanically exfoliated using the well-known tape method [1]. Although each single V_2Se_9 chains are linked by weak van der Waals interaction, we obtained a thin V_2Se_9 nanoribbon on a 300 nm SiO_2/Si substrate (see in Figure 2). Unlike typical 2D materials, an exfoliated V_2Se_9 nanoribbon shows a rough surface.

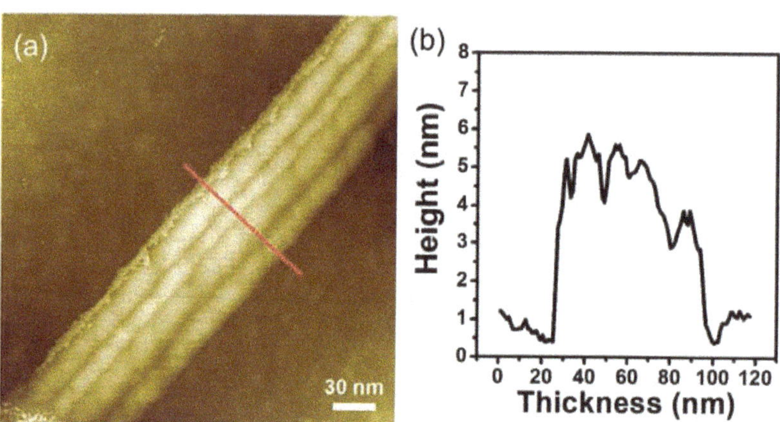

Figure 2. (**a**) Atomic force microscopy (AFM) image of the 1D V_2Se_9 flake on 300 nm SiO_2/Si substrate. (**b**) Line-profile of a V_2Se_9 flake as marked in Figure 2a.

We attempted a further delamination at the sample position using the tape, and found that some of them had been torn out (black dotted line) and that the thickness decreased from 90 to 20 nm (L1 to L1′), and from 31 to 2 nm (L2 to L2′) (see Figure 3).

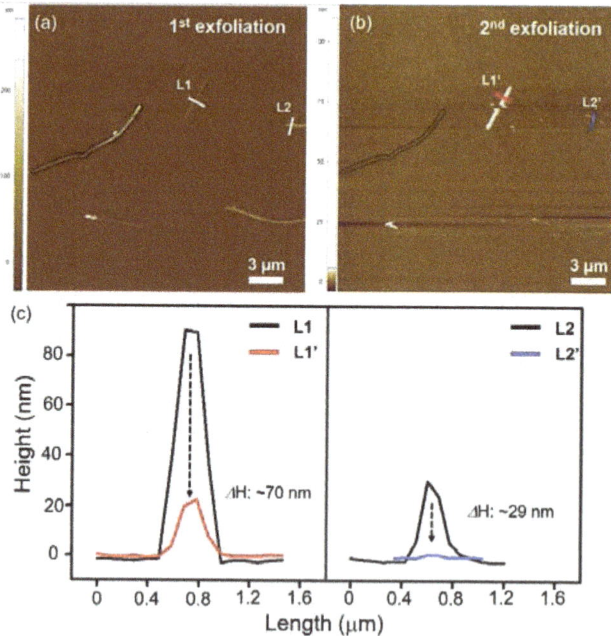

Figure 3. (**a**) AFM image of exfoliated V_2Se_9 on 300 nm SiO_2/Si substrate. (**b**) AFM image of additionally exfoliated V_2Se_9 on 300 nm SiO_2/Si substrate. (**c**) Line-profile of 1D V_2Se_9 flakes on 300 nm SiO_2/Si substrate before and after 2nd exfoliation.

Figure 4a shows the AFM image of an isolated V_2Se_9 nanoribbon on the 300 nm SiO_2/Si substrate. The nanoribbon has an atomic scale thickness and a width of approximately 20 nm (Figure 4b). Since V_2Se_9 has a bundle structure in which single chains are bonded by van der Waals forces, we expect that V_2Se_9 nanoribbons may exhibit ideal transport characteristics without degradation due to edge scattering.

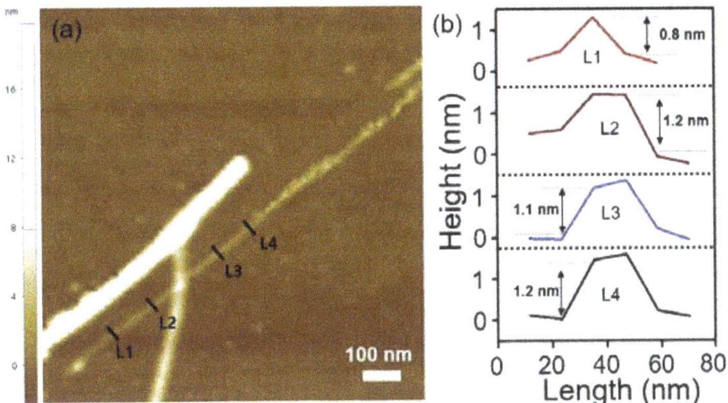

Figure 4. (**a**) AFM image of the V_2Se_9 nanoribbon on the 300 nm SiO_2/Si substrate. The inset shows an illustration of the V_2Se_9 nanoribbon. (**b**) Line-profiles of the V_2Se_9 nanoribbon as marked L1, L2, L3, and L4 in Figure 4a.

To investigate the electrical properties of V_2Se_9 flakes with a different number of layers, we performed an SKPM analysis, which is a non-destructive analytical tool that can investigate the local surface potential energy and work function by measuring the contact potential difference between the tip and the sample (V_{CPD}) [27,28]. Because the V_2Se_9 nanoribbons were on the bare Si substrate, the work function of V_2Se_9 flakes can be calculated using the following equation:

$$V_{CPD} = \frac{1}{e}\left(\varphi_t - \varphi_f\right),\tag{1}$$

$$\begin{aligned}\Delta V_{CPD} &= V_{CPD}(V_2Se_9) - V_{CPD}(substrate)\\ &= \frac{1}{e}\left(\varphi_t - \varphi_f\right) - \frac{1}{e}\left(\varphi_t - \varphi_s\right)\\ &= \frac{1}{e}\left(\varphi_s - \varphi_f\right)\end{aligned}\tag{2}$$

where φ_t, φ_s and φ_f represent the work functions of the tip, Si substrate, and V_2Se_9 flake, respectively.

As shown in Figure 5a, the surface potential energy varies with the thickness of the V_2Se_9 flakes. For example, the surface potential energy differences (Δ potential energy) between the V_2Se_9 flakes with thicknesses of 5 and 40 nm and the Si substrate were 38 and 60 mV, respectively (Figure 5b,c). A Statistical analysis of more than 27 samples shows that as the thickness of the V_2Se_9 flake is less than 25 nm, the surface potential energy difference and the work function become to decrease simultaneously (Figure 5d,e). These phenomena can be explained using an interlayer screening effect, which is also observed in typical 2D materials [27–29]. In general, the native Si oxide (e.g., SiO_x), which forms naturally on the surface of the Si wafer, has a hydrophilic property, which caused a large number of charge-trapping sites owing to the moisture in the air. Therefore, it affected the charge transfer between the V_2Se_9 flakes and the Si substrate [27]. Since the effective area of the interlayer screening effects increases with decreasing flake thickness, the surface potential difference and the work function of 25 nm thick V_2Se_9 flakes decreased from that of the bulk V_2Se_9 (See in Figure S2).

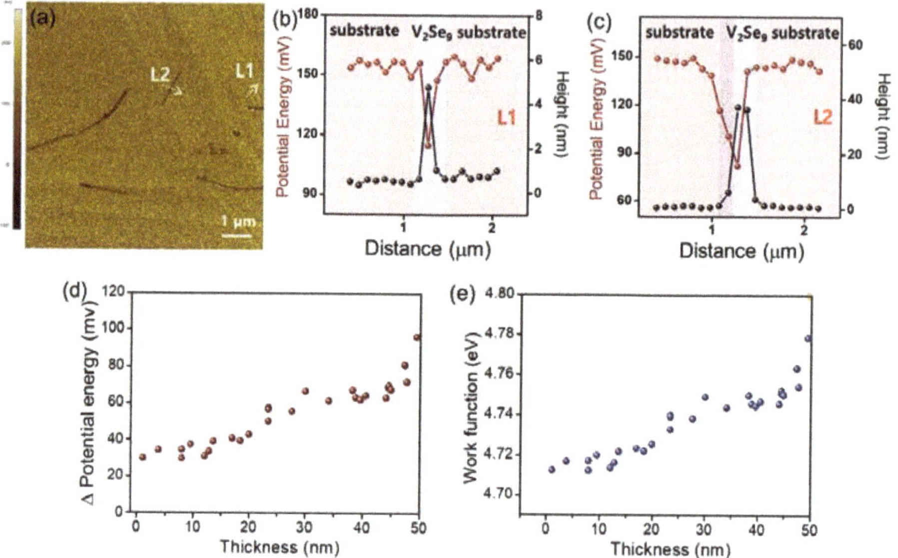

Figure 5. (**a**) Scanning Kelvin probe microscopy (SKPM) image of exfoliated 1D V_2Se_9 flakes on the Si substrate. (**b**,**c**) Height and potential energy profiles of the V_2Se_9 flakes and Si substrate as labeled in Figure 5a. (**d**,**e**) Variation in potential energy difference and work function depending as a function of thickness of V_2Se_9 flakes.

4. Conclusions

In conclusion, the mass production of the high-purity and high-crystalline 1D material V_2Se_9 crystals was successfully demonstrated using the solid-state reaction of V and Se. Through the mechanical exfoliation method, we confirmed that as-grown V_2Se_9 crystals consist of innumerable covalently bonded V_2Se_9 chains linked by the van der Waals interaction. In addition, atomic nanoribbons structures of V_2Se_9 was obtained on the 300 nm SiO_2/Si substrate. We used SKPM analysis to investigate the electrical characteristics of V_2Se_9 and established that the work function decreased with decreasing thickness of the V_2Se_9 flakes owing to the interlayer screening effect. These results will be of great help in selecting suitable metal contacts for V_2Se_9; these will have a significant influence on the overall performance. We believe that the 1D semiconductor V_2Se_9 crystal is expected to be a new family of 2D materials that will be considered essential in future device applications.

Supplementary Materials: The following are available online at http://www.mdpi.com/2079-4991/8/9/737/s1, Figure S1: Phase diagram of V-Se binary system, Figure S2: SKPM image of exfoliated 1D V_2Se_9 flake.

Author Contributions: J.-Y.C. designed the experiments, and B.J.K. and B.J.J. supported the elemental analysis. S.O. and S.C. performed the chemical reaction experiments and K.H.C., T.N. and S.H.L. support the chemical reaction experiment and K.-W.K., H.K.L., I.J.C. and J.-Y.M. supported structural analysis. H.K.Y., J.-H.L. and J.-Y.C. conceived and supervised this study and provided intellectual and technical guidance.

Funding: This work was supported by the Technology Innovation Program (or Industrial Strategic Technology Development Program) (10063400, Development of Growth and Transfer Technology for Defectless 350×350 mm^2 Single Crystalline Graphene) funded By the Ministry of Trade, Industry and Energy (MOTIE, Korea). J.H.L. acknowledges support from the Presidential Postdoctoral Fellowship Program of the National Research Foundation in Korea (2014R1A6A3A04058169).

Conflicts of Interest: The authors declare no conflict of interest.

References

1. Novoselov, K.S.; Geim, A.K.; Morozov, S.V.; Jiang, D.; Katsnelson, M.I.; Grigorieva, I.V.; Dubonos, S.V.; Firsov, A. Two-dimensional gas of massless Dirac fermions in graphene. *Nature* **2005**, *438*, 197–200. [CrossRef] [PubMed]
2. Zhang, Y.; Tan, Y.-W.; Stormer, H.L.; Kim, P. Experimental observation of the quantum Hall effect and Berry's phase in graphene. *Nature* **2005**, *438*, 201–204. [CrossRef] [PubMed]
3. Yang, H.; Heo, J.; Park, S.; Song, H.J.; Seo, D.H.; Byun, K.E.; Kim, P.; Yoo, I.; Chung, H.J.; Kim, K. Graphene barristor, a triode device with a gate-controlled Schottky barrier. *Science* **2012**, *336*, 1140–1143. [CrossRef] [PubMed]
4. Geim, A.K.; Grigorieva, I.V. Van der Waals heterostructures. *Nature* **2013**, *499*, 419–425. [CrossRef] [PubMed]
5. Lee, C.; Wei, X.; Kysar, J.W.; Hone, J. Measurement of the elastic properties and intrinsic strength of monolayer graphene. *Science* **2008**, *321*, 385–388. [CrossRef] [PubMed]
6. Son, Y.W.; Cohen, M.L.; Louie, S.G. Half-metallic graphene nanoribbons. *Nature* **2006**, *444*, 347–349. [CrossRef] [PubMed]
7. Avouris, P.; Chen, Z.; Perebeinos, V. Carbon-based electronics. *Nat. Nanotechnol.* **2007**, *2*, 605–615. [CrossRef] [PubMed]
8. Radisavljevic, B.; Radenovic, A.; Brivio, J.; Giacometti, I.V.; Kis, A. Single-layer MoS_2 transistors. *Nat. Nanotechnol.* **2011**, *6*, 147–150. [CrossRef] [PubMed]
9. Qiao, J.; Kong, X.; Hu, Z.X.; Yang, F.; Ji, W. High-mobility transport anisotropy and linear dichroism in few-layer black phosphorus. *Nat. Commun.* **2014**, *5*, 4474. [CrossRef] [PubMed]
10. Mayorov, A.S.; Gorbachev, R.V.; Morozov, S.V.; Britnell, L.; Jalil, R.; Ponomarenko, L.A.; Blake, P.; Novoselov, K.S.; Watanabe, K.; Taniguchi, T.; et al. Micrometer-scale ballistic transport in encapsulated graphene at room temperature. *Nano Lett.* **2011**, *11*, 2396–2399. [CrossRef] [PubMed]
11. Schwierz, F. Graphene transistors. *Nat. Nanotechnol.* **2010**, *5*, 487–496. [CrossRef] [PubMed]
12. Avouris, P.; Freitag, M.; Perebeinos, V. Carbon-nanotube photonics and optoelectronics. *Nat. Photonics* **2008**, *2*, 341–350. [CrossRef]
13. Wang, J.; Musameh, M. Carbon nanotube/teflon composite electrochemical sensors and biosensors. *Anal. Chem.* **2003**, *75*, 2075–2079. [CrossRef] [PubMed]
14. Chen, Z.; Augustyn, V.; Wen, J.; Zhang, Y.; Shen, M.; Dunn, B.; Lu, Y. High-performance supercapacitors based on intertwined CNT/V_2O_5 nanowire nanocomposites. *Adv. Mater.* **2011**, *23*, 791–795. [CrossRef] [PubMed]
15. Kusunoki, M.; Suzuki, T.; Honjo, C.; Hirayama, T.; Shibata, N. Selective synthesis of zigzag-type aligned carbon nanotubes on SiC (0 0 0–1) wafers. *Chem. Phys. Lett.* **2002**, *366*, 458–462. [CrossRef]
16. McCarthy, D.N.; Nicolosi, V.; Vengust, D.; Mihailovic, D.; Compagnini, G.; Blau, W.J.; Coleman, J.N. Dispersion and purification of $Mo_6S_3I_6$ nanowires in organic solvents. *J. Appl. Phys.* **2007**, *101*, 014317. [CrossRef]
17. Golden, J.H.; DiSalvo, F.J.; Fréchet, J.M.J.; Silcox, J.; Thomas, M.; Elman, J. Subnanometer-diameter wires isolated in a polymer matrix by fast polymerization. *Science* **1996**, *273*, 782–784. [CrossRef] [PubMed]
18. Messer, B.; Song, J.H.; Huang, M.; Wu, Y.; Kim, F.; Yang, P. Surfactant-induced mesoscopic assemblies of inorganic molecular chains. *Adv. Mater.* **2000**, *12*, 1526–1528. [CrossRef]
19. Heidelberg, A.; Bloeß, H.; Schultze, J.W.; Booth, C.J.; Samulski, E.T.; Boland, J.J. Electronic properties of $LiMo_3Se_3$ nanowires and Mo_3Se_3 nanowire-networks for nanoscale electronic devices. *Z. Phys. Chem.* **2003**, *217*, 573–585. [CrossRef]
20. Osterloh, F.E.; Hiramatsu, H.; Dumas, R.K.; Liu, K. $Fe_3O_4LiMo_3Se_3$ nanoparticle clusters as superparamagnetic nanocompasses. *Langmuir* **2005**, *21*, 9709–9713. [CrossRef] [PubMed]
21. Meden, A.; Kodre, A.; Gomilšek, J.P.; Arčon, I.; Vilfan, I.; Vrbanic, D.; Mrzel, A.; Mihailovic, D. Atomic and electronic structure of $Mo_6S_{9-x}I_x$ nanowires. *Nanotechnology* **2005**, *16*, 1578–1583. [CrossRef]
22. Ploscaru, M.I.; Kokalj, S.J.; Uplaznik, M.; Vengust, D.; Turk, D.; Mrzel, A.; Mihailovic, D. $Mo_6S_{9-x}I_x$ nanowire recognitive molecular-scale connectivity. *Nano Lett.* **2007**, *7*, 1445–1448. [CrossRef] [PubMed]
23. Sun, N.; McMullan, M.; Papakonstantinou, P.; Gao, H.; Zhang, X.; Mihailovic, D.; Li, M. Bioassembled nanocircuits of $Mo_6S_{9-x}I_x$ nanowires for electrochemical immunodetection of estrone hapten. *Anal. Chem.* **2008**, *80*, 3593–3597. [CrossRef] [PubMed]

24. Zhou, Y.; Wang, L.; Chen, S.; Qin, S.; Liu, X.; Chen, J.; Xue, D.J.; Luo, M.; Cao, Y.; Cheng, Y.; et al. Thin-film Sb$_2$Se$_3$ photovoltaics with oriented one-dimensional ribbons and benign grain boundaries. *Nat. Photonics* **2015**, *9*, 409–415. [CrossRef]

25. Xu, X.; Jeong, S.; Rout, C.S.; Oh, P.; Ko, M.; Kim, H.; Kim, M.G.; Cao, R.; Shin, H.S.; Cho, J. Lithium reaction mechanism and high rate capability of VS$_4$ graphene nanocomposite as an anode material for lithium batteries. *J. Mater. Chem. A* **2014**, *2*, 10847–10853. [CrossRef]

26. Britto, S.; Leskes, M.; Hua, X.; Hébert, C.A.; Shin, H.S.; Clarke, S.; Borkiewicz, O.; Chapman, K.W.; Seshadri, R.; Cho, J.; et al. Multiple Redox Modes in the Reversible Lithiation of High-Capacity, Peierls-Distorted Vanadium Sulfide. *J. Am. Chem. Soc.* **2015**, *137*, 8499–8508. [CrossRef] [PubMed]

27. Lee, N.J.; Yoo, J.W.; Choi, Y.J.; Kang, C.J.; Jeon, D.Y.; Kim, D.C.; Seo, S.; Chung, H.J. The interlayer screening effect of graphene sheets investigated by Kelvin probe force microscopy. *Appl. Phys. Lett.* **2009**, *95*, 222107. [CrossRef]

28. Choi, S.H.; Shaolin, Z.; Yang, W. Layer-number-dependent work function of MoS$_2$ nanoflakes. *J. Korean Phys. Soc.* **2014**, *64*, 1550–1555. [CrossRef]

29. Li, Y.; Xu, C.-Y.; Zhen, L. Surface potential and interlayer screening effects of few-layer MoS$_2$ nanoflakes. *Appl. Phys. Lett.* **2013**, *102*, 143110. [CrossRef]

nanomaterials  

Article

Feasible Route for a Large Area Few-Layer MoS$_2$ with Magnetron Sputtering

Wei Zhong [1,†], Sunbin Deng [2,†], Kai Wang [2], Guijun Li [2], Guoyuan Li [1], Rongsheng Chen [1,2,*] and Hoi-Sing Kwok [2]

[1] School of Electronic and Information Engineering, South China University of Technology, Guangzhou 510640, China; zwnice@163.com (W.Z.); phgyli@scut.edu.cn (G.L.)
[2] State Key Laboratory on Advanced Displays and Optoelectronics Technologies, Department of Electronic and Computer Engineering, The Hong Kong University of Science and Technology, Hong Kong, China; sdengaa@connect.ust.hk (S.D.); kaiwjnu@163.com (K.W.); gliad@connect.ust.hk (G.L.); eekwok@ust.hk (H.-S.K.)
* Correspondence: rschen@connect.ust.hk; Tel.: +86-20-8711-1435
† These authors have contributed equally to this work.

Received: 20 June 2018; Accepted: 30 July 2018; Published: 3 August 2018

Abstract: In this article, we report continuous and large-area molybdenum disulfide (MoS$_2$) growth on a SiO$_2$/Si substrate by radio frequency magnetron sputtering (RFMS) combined with sulfurization. The MoS$_2$ film was synthesized using a two-step method. In the first step, a thin MoS$_2$ film was deposited by radio frequency (RF) magnetron sputtering at 400 °C with different sputtering powers. Following, the as-sputtered MoS$_2$ film was further subjected to the sulfurization process at 600 °C for 60 min. Sputtering combined with sulfurization is a viable route for large-area few-layer MoS$_2$ by controlling the radio-frequency magnetron sputtering power. A relatively simple growth strategy is demonstrated here that simultaneously enhances thin film quality physically and chemically. Few-layers of MoS$_2$ are established using Raman spectroscopy, X-ray diffractometer, high-resolution field emission transmission electron microscope, and X-ray photoelectron spectroscopy measurements. Spectroscopic and microscopic results reveal that these MoS$_2$ layers are of low disorder and well crystallized. Moreover, high quality few-layered MoS$_2$ on a large-area can be achieved by controlling the radio-frequency magnetron sputtering power.

Keywords: few-layer MoS$_2$; magnetron sputtering; magnetron sputtering power; raman spectroscopy; disorder

1. Introduction

The emergence of monolayer graphene [1,2] and transition metal dichalcogenides (TMDs) [3,4] has inspired a series of high-profile discoveries in the electronic and optoelectronic fields [5–7], and has initiated potentially new areas [8,9]. Thus, two-dimensional (2D) materials have recently been intensively studied. In the context of 2D TMDs, molybdenum disulfide (MoS$_2$) is one of the attractive embodiments due to its stable form in few- and single-layers [10,11] as well as its desirable electrical and optical properties [12]. Few- and single-layer MoS$_2$ is firstly obtained via top-down mechanical stripping techniques [13], which are commonly used for graphene exfoliation. Although this method of exfoliation and dip coating [14–17] has its own the advantages to achieve high-quality 2D materials, none of these are proper solutions for radical large-scale commercial manufacturing, where the mass-producible growth of large-area, continuous, and high-quality 2D MoS$_2$ thin films on dielectrics is a pre-requisite. From this point of view, the bottom-up strategies for thin film growth, including chemical vapor deposition (CVD) and physical vapor deposition (PVD), are better choices when compared with the top-down methods mentioned above. CVD methods have already been successfully demonstrated [18–20], but the control of thin film thickness, purity and uniformity on a large scale needs to be further enhanced [21]. On the other hand, PVD methods, especially magnetron sputtering, have been broadly employed in large-scale commercial manufacturing at low cost and with easy control. However, the exploration of 2D MoS_2 thin film growth using magnetron sputtering technology is quite insufficient [21]. In recent years, there have been few attempts at sputtering techniques for the growth of MOS_2 thin films. Muratore et al. [22], Kaindl et al. [23], and Samassekou et al. [24] reported the synthesis of continuous few-layer MoS_2 by sputtering method using a MoS_2 target, and the sputtered MoS_2 films was annealed in an argon atmosphere. Tao et al. [21] and Santoni et al. [25] reported MoS_2 film using Mo target sputtered in vaporized sulfur ambient. However, the reported films either are relatively thick or have poor crystal quality and optical properties [21–25]. The main obstacles for large-area sputtered high quality MoS_2 films are possibly located in the difficulty of disorder control during thin film deposition, and the lack of metrics to evaluate the deposited thin films.

Studies over the past decade have shown that Raman spectroscopy has historically played an important role in the structural characterization of graphitic materials [26–28] and has also become a powerful tool to understand the effects of process parameters on the production of high quality graphene by monitoring changes in disordered peaks [28–32]. Recently, Raman spectroscopy has also been used to study the effects of disorder on the MoS_2 [33]. On the other hand, since many scholars have previously investigated the crystallinity of MoS_2 thin film by high-temperature vulcanization or deposition on different substrates [21,34,35], few studies have investigated the effect of sputtering power on the MoS_2 thin film. Therefore, in this work, we use a Raman spectroscopy approach to describe the quality and to study the effect of RF power on the deposition of the large-scale few-layer MoS_2 films. We also investigated the crystalline structure of the films through an X-ray diffractometer (XRD) and a high-resolution field emission transmission electron microscope (HRTEM). The binding energies of Mo in the MoS_2 film grown by radio frequency magnetron sputtering (RFMS) were further analyzed by X-ray photoelectron spectroscopy (XPS). For precise detection, the surface 5-nm-thick thin films were etched using Ar ions before XPS characterization.

2. Materials and Methods

The large-scale few layer MoS_2 thin films were deposited using the RFMS technique. Firstly, both silicon substrates coated with thermally grown SiO_2 and glass substrates (Eagle 2000, Corning) were ultrasonically cleaned in acetone and then isopropanol (IPA). After being rinsed in deionized (DI) water and dried, the substrates were loaded into the chamber of an RFMS system (AJA International Inc., Scituate, MA, USA) and heated to 400 °C. The distance between the substrate and the MoS_2 target (99.99%, 2 Inc., Plasmaterials Inc., Livermore, CA, USA) was 10 cm. In order to suppress the influence of oxygen and moisture on the deposited thin films, the base pressure of the chamber should be pumped down as low as possible. In this work, the value was 2.67×10^{-5} Pa. During deposition, Ar gas (99.999%) was allowed to flow into the chamber with a stable flow rate of 20 sccm, and the working pressure of the chamber was maintained at 3×10^{-3} Torr. The RF sputtering power applied to the MoS_2 target was varied from 10 W to 150 W in order to investigate the relationship between the large-scale few layer MoS_2 growth and RF power. The deposition time was dependent on the required thickness of the thin films and the thickness of the film is maintained at 15 nm, the thickness of the thin films is confirmed by HRTEM (High-Resolution Transmission Electron Microscopy). When the RFMS process was completed, the heater and the Ar gas pipeline were switched off, and the substrates were cooled down to room temperature naturally. After loading out of the chamber, the SiO_2/Si substrates and glass substrates with as-deposited MoS_2 thin film were immediately subjected to post-annealing in a sulfurization atmosphere for 60 min at 600 °C to enhance thin film quality physically and chemically (Figure 1).

In order to identify the layered structure and obtain the disorder information, the deposited MoS_2 thin films were analyzed using Raman spectroscopy (Renishaw invia RE04, 514 nm Ar laser with a 1 μm spot size, Renishaw plc, Gloucestershire, United Kingdom). For the crystalline phase characterizations of the thin films, an X-ray diffractometer (XRD, Empyrean, PANalytical, Almelo, The Netherlands) with Cu Kα radiation was used. It was operated in thin-film mode, and the angle between the X-ray and the thin film surface was fixed at 0.5 degrees. Furthermore, a high-resolution transmission electron microscopy (HRTEM, JEM-2010HR, JEOL, Tokyo, Japan) was applied to identify the number of layers and atomic structure of the MoS_2 film. To analyze the material composition of the deposited thin films as well as the chemical environment of the atoms in the thin films, X-ray photoelectron spectroscopy (XPS) measurement was conducted on the Physical Electronics 5600 multi-technique system (Physical Electronics Inc., Chanhassen, MN, USA).

3. Results and Discussion

Figure 1 shows an image of MoS_2 thin layer grown on glass substrates with different deposition time (15 s, 30 s, and 45 s). The deposited MoS_2 layer was light gray, and after annealing under a sulfur atmosphere, the MoS_2 layer was pale yellow and was found to have specular reflection of ambient light. This result is consistent with previous reports [21]. The size of the synthesized films is limited by the dimensions of our sample heating holder.

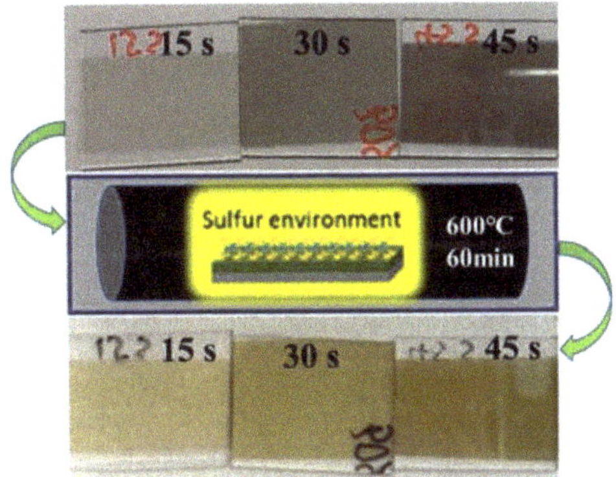

Figure 1. This post-deposition annealing treatment was performed to further enhance crystalline quality in as-sputtered MoS_2 on glass substrates under sulfur environment.

Figure 2 shows the Raman spectra of the MoS_2 thin films on SiO_2/Si substrates deposited under different RF powers of 10 W, 80 W, 120 W and 150 W, respectively. From Figure 2, it can be seen that all of the thin films exhibit two specific Raman characteristic peaks of MoS_2, namely the in-plane (E_{2g}^1) peak at ~381 cm^{-1} and the out-of-plane (A_{1g}) peak at ~405 cm^{-1}. Their peak positions herein are quite consistent with the results in other reports [24,36,37], which the E_{2g}^1 and A_{1g} mode peaks appear at ~381.5 cm^{-1} and ~404.8 cm^{-1}. Besides this, a relatively small LA(M) peak (a defect peak generated by LA phonons at the M point in the Brillouin zone) is also observed at ~227 cm^{-1}. Theoretically, the E_{2g}^1 mode corresponds to the S and Mo atoms oscillating in antiphase parallel to the crystal plane, while the A_{1g} mode corresponds to the S atoms oscillating in antiphase out-of-plane, as shown in the insets of Figure 2 [29,38]. In addition to the absolute positions of these two peaks, the frequency difference (Δk) between the A_{1g} peak and the E_{2g}^1 peak is a good indicator of the layer number in MoS_2 thin films. In this work, Δk is around 24 cm^{-1}, which is larger than that in monolayer MoS_2 thin films (~18–19 cm^{-1}) [19,33,39], but smaller than the typical value in bulk MoS_2 (~26 cm^{-1}) [37,38]. This indicates the existence of few-layer MoS_2 [36]. Table 1 lists the peak positions corresponding to the E_{2g}^1 and A_{1g} mode as well as the Δk values for all samples under various RF sputtering powers in this work. With the continuous increase of RF power from 10 W to 150 W, the Δk values remain in the vicinity of 24 cm^{-1}. This means that all of the thin films under different RF powers are MoS_2 with a few layers [36,37]. Another indicator of film quality is the full width at half maxima (FWHM) of the observed vibration modes. FMHM values for the sputtered FL-MoS$_2$ film with different RF powers is compared in Table 1. In general, higher FMHM values mean more disorder [24,25]. From the Table 1, it can be seen that the MoS$_2$ films with RF powers of 120 W has the lowest FMHM values, meaning it has the least disorder. It also can be seen that the A$_{1g}$ peak and the E$_{2g}^1$ peak all have a blue shift as the RF powers increased from 10 W to 120 W and a red shift as the RF powers increased from 120 W to 150 W. The red shift is attributed to the high RF power, resulting in an increase in the residual-stress of the film [40].While the blue shift is attributed to O$_2$-doping of MoS$_2$, which will be shown on the XPS result [41].

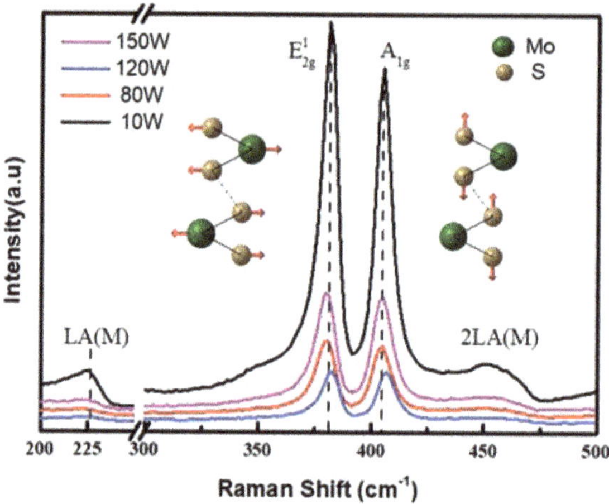

Figure 2. Raman spectra of the MoS$_2$ thin films deposited on SiO$_2$/Si substrates under different radio frequency (RF) powers. The insets illustrate the oscillating mode of the E$_{2g}^1$ and A$_{1g}$ peak.

According to Raman fundamental selection rules, only phonons with wave vector q $\cong$ 0 are Raman active around the center of the Brillouin zone. However, this rule will be broken by defects, which cause the appearance of peaks away from the zone center [33,42]. Monitoring the evolution of disorder-related sub-peaks in the Raman spectrum enables us to understand the effects of process parameters and has allowed great strides to be made in the CVD preparation of high quality graphene [29,43,44]. Similarly, in Figure 2, apart from two major peaks with regard to the basic E$_{2g}^1$ and A$_{1g}$ vibration modes, several sub-peaks related to the defects can also be observed. Among them, the sub-peak at 227 cm^{-1} is the most intense, which is attributed to the longitudinal phonon branch at the point M (LA (M)) of the Brillouin region [33,45]. Therefore, this sub-peak is able to form a very clear marker for disorder in the system, especially for the few- and single-layer MoS$_2$. The intensity ratio of the LA(M) sub-peak to A$_{1g}$ peak is plotted as a function of RF power in Figure 3. It can be clearly observed that the intensity ratio reaches a minimum when the RF power climbs to 120 W, indicating the improvement of thin film quality [46]. In general, higher RF power could assist with the formation of crystalline films with lower disorder (namely, lower LA(M)/ A$_{1g}$ intensity ratio), but excessively high RF power is not welcomed. For instance, the LA(M)/A$_{1g}$ intensity ratio rises when the RF power increases from 120 W to 150 W. It reveals the increase of disorder in the deposited MoS$_2$ thin films. A possible explanation for this phenomenon could be the increased defect generation caused by ion bombardment under over-high RF power.

Table 1. The E^1_{2g}- and A_{1g}-related Raman peak information of MoS_2 thin films deposited using radio frequency magnetron sputtering (RFMS) under various RF powers.

RF Power (W)	A_{1g} (cm^{-1})	E^1_{2g} (cm^{-1})	Δk (A_{1g}-E^1_{2g}) (cm^{-1})	Full Width at Half-Maximum (cm^{-1})		LA(M) to A_{1g} Peak Intensity Ration
				A_{1g}	E^1_{2g}	
10	405.2 ± 0.1	381.2 ± 0.1	24.0 ± 0.02	10.64 ± 0.46	10.87 ± 0.56	0.219 ± 0.010
80	405.4 ± 0.4	381.4 ± 0.4	24.0 ± 0.01	10.17 ± 0.34	10.31 ± 0.21	0.195 ± 0.004
120	407.0 ± 0.1	383.0 ± 0.1	24.0 ± 0.02	9.55 ± 0.02	9.57 ± 0.20	0.180 ± 0.003
150	403.7 ± 0.4	379.7 ± 0.4	24.0 ± 0.02	10.49 ± 0.13	10.56 ± 0.02	0.204 ± 0.006

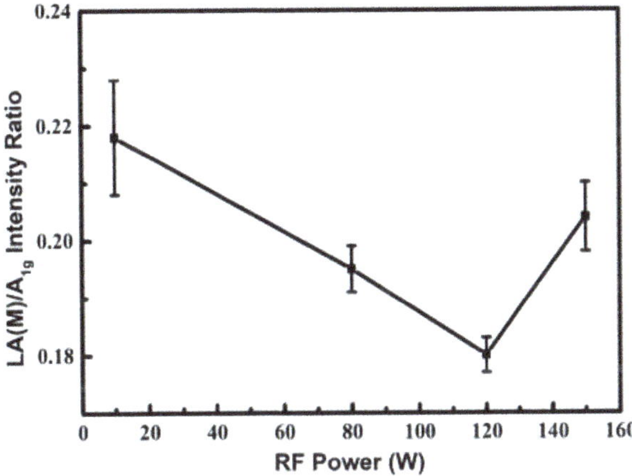

Figure 3. LA(M) to A_{1g} peak intensity ratio of the MoS_2 films with different RF powers deposited on SiO_2/Si substrates.

The crystalline information of the samples was characterized using XRD. The X-ray diffraction patterns of MoS_2 thin films on SiO_2/Si substrates under different RF powers are shown in Figure 4. The MoS_2 thin films deposited under different RF powers all exhibit three obvious diffraction peaks, which are located at 14.0°, 21.6°, and 51.5° respectively. The two weak diffraction peaks at 21.6° and 51.5° correspond to the SiO_2 (JCPDS: 27-0605) (111) and (400) planes, respectively. The strong diffraction peak at 14.0° is an indicator of the MoS_2 (JCPDS: 37-1492) (002) plane. For the MoS_2 thin films deposited under an RF power of 10 W, the broad and weak diffraction peak at 14.0° indicates the amorphous structure of the film. However, when the RF power increases to 80 W, 120 W and then 150 W, all of the MoS_2 thin films exhibit a strong and narrow diffraction peak at 14.0°, which indicate the well crystallization of MoS_2 thin films. Since the (002) plane is parallel to the surface of the substrates, the deposited MoS_2 thin films using the RFMS technique under an RF power of 80 W, 120 W and 150 W could grow along the c-axis. Meanwhile, the exclusive diffraction peak also suggests the thin films are highly oriented. These properties are quite helpful for the formation of stacked microstructures in MoS_2 thin films. Besides this, it should be noted that the intensity of the (002) peak for MoS_2 thin films under an RF power of 150 W is lower compared to the other two samples. This is possibly related to the degradation of crystalline quality under such high RF powers [47]. At the same time, we also noticed that under the RF power of 150 W, the sample showed a weak diffraction peak at 28.4°, which corresponds to the Si (JCPDS: 27-1402) (100) plane.

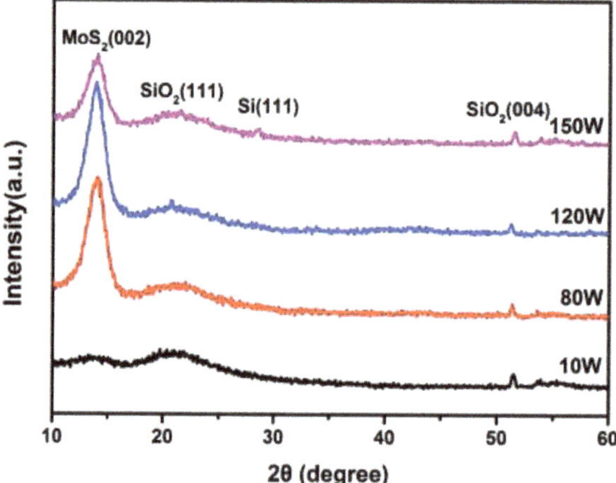

Figure 4. X-ray diffraction patterns of the MoS$_2$ thin films on SiO$_2$/Si substrates under different RF sputtering powers.

To further elucidate the crystalline structure, the MoS$_2$ thin film deposited by 120 W RF power was transferred onto a lacey copper grid for HRTEM characterizations. Figure 5a presents its cross-sectional HRTEM image, it can be seen that there are about 20 layers of the MoS$_2$ thin films on the SiO$_2$/Si substrate, and the interlayer spacing (0.68 nm) of the MoS$_2$ thin films is consistent with the previous results [21]. Moreover, it is also verified that the deposited MoS$_2$ thin films are stacked a few layers in parallel, which is in agreement with the results extracted from the Raman spectra above. As far as the typical high resolution TEM image in Figure 5b is concerned, the first-order diffraction spots of the FFT image (inset of Figure 5b) on a selected area are shown according to a regular hexagonal symmetry, thus indicating the presence of the MoS$_2$ thin film made of single crystal domains (without Moireé patterns). Moreover, HRTEM images of the selected area in Figure 5b after FFT filtering are shown in Figure 5c. The IFFT-filtered image (Figure 5c) shows a regular honeycomb pattern due to the atomic arrangements of the Mo and S atoms. In Figure 5d, the calculated profile along the selected direction in Figure 5c is drawn, which reveals the (004) plane of MoS$_2$ with a lattice spacing of 0.31 nm that is observed in Figure 5c. Figure 5e is the magnified image of the square-surrounded region in Figure 5c. The periodic atom arrangement for Mo confirms that the MoS$_2$ thin films deposited under an RF power of 120 W own a crystalline structure. This is also consistent with the XRD results in Figure 4.

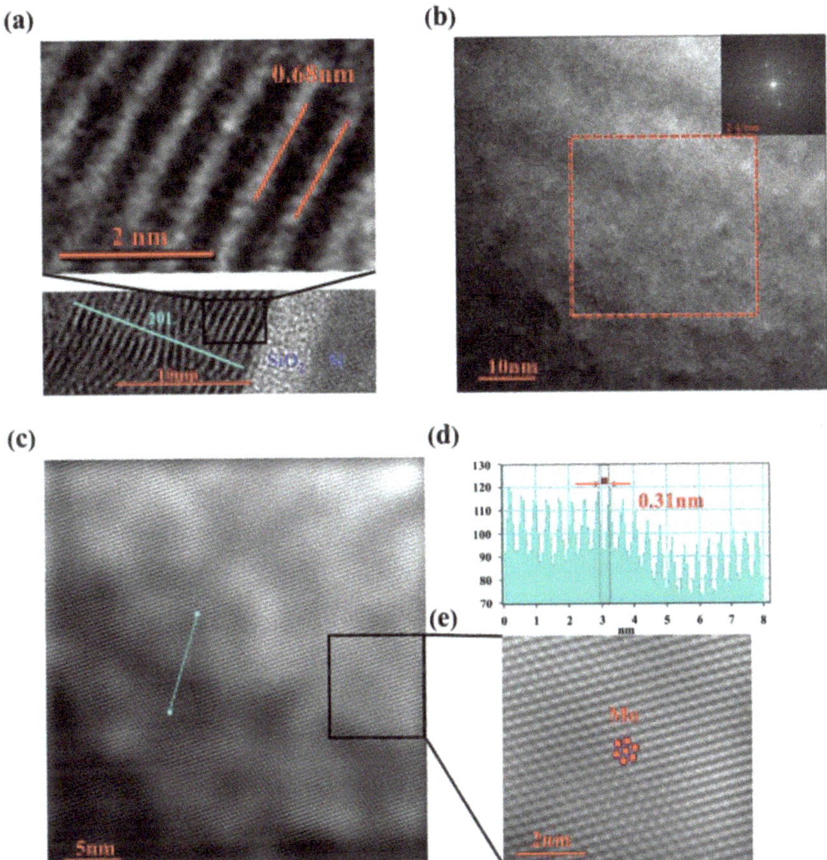

Figure 5. (**a**) Cross-sectional high-resolution field emission transmission electron microscope (HRTEM) image of samples deposited under an RF power of 120 W. (**b**) High resolution TEM image of the MoS$_2$ film deposited by 120 W RF power transferred onto a lacey carbon grid. The inset shows the fast Fourier transformation (FFT) image corresponding to the TEM image selected area of a portion of (**b**), showing the hexagonal symmetry of the MoS$_2$ structure. (**c**) Inverse FFT images corresponding to the TEM image selected area of a portion of (**b**). (**d**) Atomic spacing along the selected direction of the basal plane. (**e**) Zoom-in image of the area highlighted in (**c**). The hexagonal structure formed by Mo atoms is indicated.

XPS was used to analyze the chemical environment of Mo in the MoS$_2$ thin films, and to detect any impurity (particularly oxygen) involvement during preparation. The high-resolution XPS spectra of the Mo 3d and S 2s region are shown in Figure 6. Since the peaks of Mo 3d and S 2s are too close to distinguish, in order to obtain the chemical environment information of the Mo species, both the S 2s and Mo 3d spectra are taken into account. Moreover, since the main Mo doublet peak signals of the thin films under an RF power of 10 W are so weak that they reach noise level, the analysis in terms of such samples is not reliable. Hence, only the data with an RF power of 80 W, 120 W and 150 W have been fitted by a 20% Lorentzian–Gaussian ration fit and the according results together with the Shirley background are presented in Figure 6b–d. According to [35], the two peaks at 229.1 eV and 232.2 eV for the MoS$_2$ thin film are attributed to the doublet Mo 3d$_{5/2}$ and Mo 3d$_{3/2}$ orbitals, respectively. Meanwhile, the fitting shows that there is a second characteristic at the lower binding energy. For species originating from lower binding energy (BE) Mo3d peaks, the lower BE Mo species is Mo still associated with the Mo-S lattice, or it reflects a single amorphous MoS_x phase, where Mo has a different number of nearest neighbor S atoms [48,49]. According to [25], the lower BE Mo species can be assigned to zero-valent Mo (Mo(0)) occurring in small aggregates dispersed in a MoS_2 matrix. From the principle of sputter coating, we can know that the presence of Mo(0) cannot be avoided, and the purpose of annealing under a sulfur atmosphere is to convert Mo(0) to Mo(IV), which can reduce disorder and defect formation. However, when the film is etched by an ion beam, there is also a chemical shift of its binding energy toward smaller values [25,35]. Therefore, compared to XPS characterization, the advantages of strong non-destructive characterization of Raman spectra are even more pronounced. For accurate detection, 5-nano-thin-thick-surface films should be etched with Ar ions prior to XPS characterization. However, this will lead to sample destruction and will have a certain impact on the test results. Although this changed the chemical state of the Mo atom, all the samples were processed in the same way, and the overall trend of its variation with sputtering power did not change. The relative ratio of Mo species calculated from the deconvolved Mo 3d spectrum results is shown in Figure 7. The Mo(IV) fraction is defined as the area under the Mo(IV) peak divided by the total Mo peak area. It can clearly be seen that the Mo (IV) fraction reaches a maximum when the RF power is increased to 120 W, indicating the improvement of thin film quality. In addition, smaller peaks at 230.2 eV and 233.3 eV are assigned to Mo^{5+} [50], indicating the presence of chemisorbed oxygen at sulfur vacancies [51,52] or sub-stoichiometric oxides MoO_x [53]. The origin of MoO_x is probably due to the sulfurization process. We believe that MoO_x has probably formed during the deposition and/or during the sulfurization process at 600 °C by reaction with the oxygen diffused from the SiO_2/Si substrate. As shown in Figure 7, the Mo (V) fraction is around 0.1, 0.094 and 0.102 for an RF power of 80 W, 120 W, and 150 W, respectively. For different RF powers, the Mo(V) fraction is basically the same, indicating that the power change has no effect on the Mo(V) fraction.

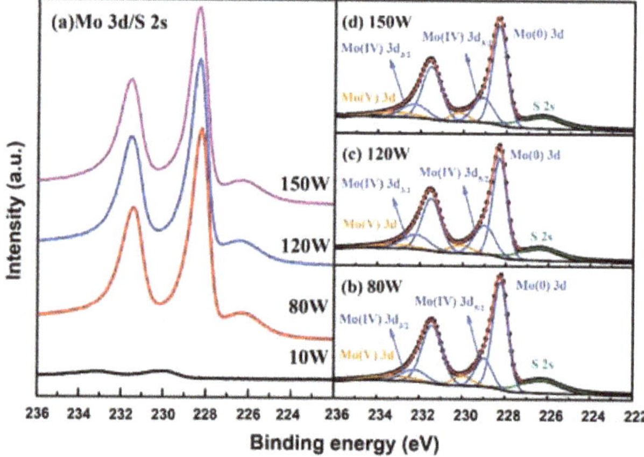

Figure 6. High-resolution X-ray photoelectron spectroscopy (XPS) spectra of (**a**) Mo 3d/S 2s, and (**b**), (**c**), and (**d**) Mo3d core-level spectra for an RF power of 80 W, 120 W, and 150 W, respectively. The background is shown with a black line at the bottom. The black dots represent the raw data. The red line is the total least-squares fit. The orange lines indicate the Mo(V) 3d components. The blue lines are the Mo3d components linked to Mo (0) and Mo (IV), the Mo (0) are the lower BE doublet. The olive lines are the S2s components. (See text for more explanation).

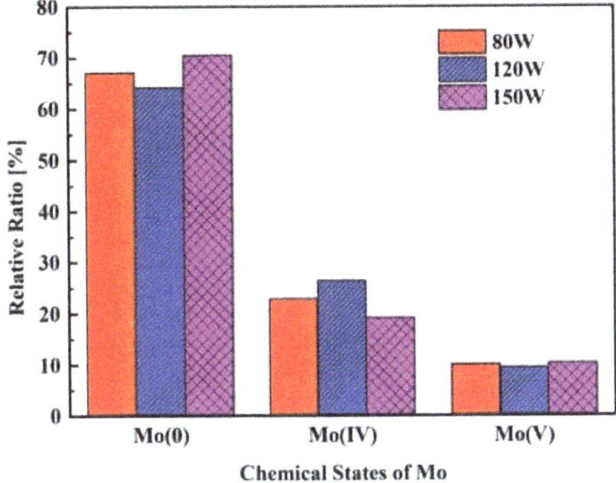

Figure 7. Relative ratio of Mo species with various chemical states at different RF powers.

4. Conclusions

Few-layer MoS_2 thin film deposition on large-area thermally oxidized silicon substrates was demonstrated using the RFMS technique. Raman analysis verified the achievement of few-layer MoS_2 thin films. Meanwhile, it was proposed that the disorder inside the thin films could be monitored using Raman spectra, and controlled by adjusting the RF sputtering power. Furthermore, the XRD spectra and cross-sectional TEM images confirmed the high quality of few-layer MoS_2 thin films, implying that RFMS was suitable for layered MoS_2 growth. Additionally, the XPS characterizations on RFMS grown few-layer MoS_2 thin films revealed the RF power has a great effect on the binding energies of Mo atoms. Our work illustrates that sputtering combined with sulfurization is a viable route for the high quality of large-area few-layer MoS_2 by controlling the radio-frequency magnetron sputtering power.

Author Contributions: S.D. and K.W. conceived and designed the experiments; S.D. and K.W. performed the experiments; W.Z. and R.C. analyzed the data; S.D., K.W., W.Z. and G.L. (Guijun Li). contributed materials/analysis tools; W.Z., S.D. and R.C. wrote the paper; G.L. (Guoyuan Li) and H.-S.K. provided advice about the content and the structure of this work.

Funding: This research was funded by the National Natural Science Foundation of China under Grant 61604057, in part by the Partner State Key Laboratory on Advanced Displays and Optoelectronics Technologies under Grant ITC-PSKL12EG02, in part by the Science and Technology Program of Guangdong Province under Grant 2017A010101010, and in part by the Science and Technology Program of Guangdong Province under Grant No. 201807010098.

Conflicts of Interest: The authors declare no conflict of interest.

References

1. Guinea, F.; Peres, N.M.R.; Novoselov, K.S.; Geim, A.K.; Castro Neto, A.H. The electronic properties of graphene. *Rev. Mod. Phys.* **2009**, *81*, 109–162.
2. Geim, A.K.; Novoselov, K.S. The rise of graphene. *Nat. Mater.* **2007**, *6*, 183–191. [CrossRef] [PubMed]
3. Miró, P.; Audiffred, M.; Heine, T. An atlas of two-dimensional materials. *Chem. Soc. Rev.* **2014**, *43*, 6537–6554. [CrossRef] [PubMed]
4. Kuc, A. Low-dimensional transition-metal dichalcogenides. In *Chemical Modelling*; Royal Society of Chemistry: London, UK, 2014; Volume 11, pp. 1–29.
5. Fiori, G.; Bonaccorso, F.; Iannaccone, G.; Palacios, T.; Neumaier, D.; Seabaugh, A.; Banerjee, S.K.; Colombo, L. Electronics based on two-dimensional materials. *Nat. Nanotechnol.* **2014**, *9*, 768–779. [CrossRef] [PubMed]
6. Wang, Q.H.; Kalantar-Zadeh, K.; Kis, A.; Coleman, J.N.; Strano, M.S. Electronics and optoelectronics of two-dimensional transition metal dichalcogenides. *Nat. Nanotechnol.* **2012**, *7*, 699–712. [CrossRef] [PubMed]

7. Kang, K.; Xie, S.; Huang, L.; Han, Y.; Huang, P.Y.; Mak, K.F.; Kim, C.J.; Muller, D.; Park, J. High-mobility three-atom-thick semiconducting films with wafer-scale homogeneity. *Nature* **2015**, *520*, 656–660. [CrossRef] [PubMed]

8. Zeng, H.; Dai, J.; Yao, W.; Xiao, D.; Cui, X. Valley polarization in MoS_2 monolayers by optical pumping. *Nat. Nanotechnol.* **2012**, *7*, 490–493. [CrossRef] [PubMed]

9. Mak, K.F.; He, K.; Shan, J.; Heinz, T.F. Control of valley polarization in monolayer MoS_2 by optical helicity. *Nat. Nanotechnol.* **2012**, *7*, 494–498. [CrossRef] [PubMed]

10. Novoselov, K.S.; Jiang, D.; Schedin, F.; Booth, T.J.; Khotkevich, V.V.; Morozov, S.V.; Geim, A.K. Two-dimensional atomic crystals. *Proc. Natl. Acad. Sci. USA* **2005**, *102*, 10451–10453. [CrossRef] [PubMed]

11. Mak, K.F.; Lee, C.; Hone, J.; Shan, J.; Heinz, T.F. Atomically thin MoS_2: A new direct-gap semiconductor. *Phys. Rev. Lett.* **2010**, *105*, 136805. [CrossRef] [PubMed]

12. Lin, Z.; Mccreary, A.; Briggs, N.; Subramanian, S.; Zhang, K.; Sun, Y.; Li, X.; Borys, N.J.; Yuan, H.; Fullerton-Shirey, S.K.; et al. 2D materials advances: From large scale synthesis and controlled heterostructures to improved characterization techniques, defects and applications. *2D Mater.* **2016**, *3*, 042001. [CrossRef]

13. Radisavljevic, B.; Radenovic, A.; Brivio, J.; Giacometti, V.; Kis, A. Single-layer MoS_2 transistors. *Nat. Nanotechnol.* **2011**, *6*, 147–150. [CrossRef] [PubMed]

14. Li, Y.; Xu, C.Y.; Hu, P.A.; Zhen, L. Carrier Control of MoS_2 Nanoflakes by Functional Self-Assembled Monolayers. *ACS Nano* **2013**, *7*, 7795–7804. [CrossRef] [PubMed]

15. Sundaram, R.S.; Engel, M.; Lombardo, A.; Krupke, R.; Ferrari, A.C.; Avouris, P.; Steiner, M. Electroluminescence in single layer MoS_2. *Nano Lett.* **2013**, *13*, 1416–1421. [CrossRef] [PubMed]

16. Das, S.; Chen, H.Y.; Penumatcha, A.V.; Appenzeller, J. High performance multilayer MoS_2 transistors with scandium contacts. *Nano Lett.* **2013**, *13*, 100–105. [CrossRef] [PubMed]

17. Sik Hwang, W.; Remskar, M.; Yan, R.; Kosel, T.; Kyung Park, J.; Cho, B.J.; Haensch, W.; Grace Xing, H.; Seabaugh, A.; Jena, D. Comparative study of chemically synthesized and exfoliated multilayer MoS_2 field-effect transistors. *Appl. Phys. Lett.* **2013**, *102*, 043116. [CrossRef]

18. Elibol, K.; Susi, T.; M, O.B.; Bayer, B.C.; Pennycook, T.J.; Mcevoy, N.; Duesberg, G.S.; Meyer, J.C.; Kotakoski, J. Grain boundary-mediated nanopores in molybdenum disulfide grown by chemical vapor deposition. *Nanoscale* **2016**, *9*, 1591–1598. [CrossRef] [PubMed]

19. Lee, Y.H.; Zhang, X.Q.; Zhang, W.; Chang, M.T.; Lin, C.T.; Chang, K.D.; Yu, Y.C.; Wang, J.T.; Chang, C.S.; Li, L.J.; et al. Synthesis of large-area MoS_2 atomic layers with chemical vapor deposition. *Adv. Mater.* **2012**, *24*, 2320–2325. [CrossRef] [PubMed]

20. Zhan, Y.; Liu, Z.; Najmaei, S.; Ajayan, P.M.; Lou, J. Large Area Vapor Phase Growth and Characterization of MoS_2 Atomic Layers on SiO_2 Substrate. *Small* **2012**, *8*, 966–971. [CrossRef] [PubMed]

21. Tao, J.; Chai, J.; Lu, X.; Wong, L.M.; Wong, T.I.; Pan, J.; Xiong, Q.; Chi, D.; Wang, S. Growth of wafer-scale MoS_2 monolayer by magnetron sputtering. *Nanoscale* **2015**, *7*, 2497–2503. [CrossRef] [PubMed]

22. Muratore, C.; Hu, J.J.; Wang, B.; Haque, M.A. Continuous ultra-thin MoS_2 films grown by low-temperature physical vapor deposition. *Appl. Phys. Lett.* **2014**, *104*, 261604. [CrossRef]

23. Kaindl, R.; Bayer, B.C.; Resel, R.; Muller, T.; Skalakova, V.; Habler, G.; Abart, R.; Cherevan, A.S.; Eder, D.; Blatter, M.; et al. Growth, structure and stability of sputter-deposited MoS_2 thin films. *Beilstein J. Nanotechnol.* **2017**, *8*, 1115–1126. [CrossRef] [PubMed]

24. Samassekou, H.; Alkabsh, A.; Wasala, M.; Eaton, M.; Walber, A.; Walker, A.; Pitkanen, O.; Kordas, K.; Talapatra, S.; Jayasekera, T.; et al. Viable route towards large-area 2D MoS_2 using magnetron sputtering. *2D Mater.* **2017**, *4*, 021002. [CrossRef]

25. Santoni, A.; Rondino, F.; Malerba, C.; Valentini, M.; Mittiga, A. Electronic structure of Ar^+ ion-sputtered thin-film MoS_2: A XPS and IPES study. *Appl. Surf. Sci.* **2017**, *392*, 795–800. [CrossRef]

26. Pimenta, M.A.; Neves, B.R.A.; Medeiros-Ribeiro, G.; Enoki, T.; Kobayashi, Y.; Takai, K.; Fukui, K.; Dresselhaus, M.S.; Saito, R.; Jorio, A.; et al. Anisotropy of the Raman Spectra of Nanographite Ribbons. *Phys. Rev. Lett.* **2004**, *93*, 047403.

27. Wilhelm, H.; Lelaurain, M.; Mcrae, E.; Humbert, B. Raman spectroscopic studies on well-defined carbonaceous materials of strong two-dimensional character. *J. Appl. Phys.* **1998**, *84*, 6552–6558. [CrossRef]

28. Pimenta, M.A.; Dresselhaus, G.; Dresselhaus, M.S.; Cançado, L.G.; Jorio, A.; Saito, R. Studying disorder in graphite-based systems by Raman spectroscopy. *Phys. Chem. Chem. Phys.* **2007**, *9*, 1276–1290. [CrossRef] [PubMed]

29. Ferrari, A.C.; Basko, D.M. Raman spectroscopy as a versatile tool for studying the properties of graphene. *Nat. Nanotechnol.* **2013**, *8*, 235–246. [CrossRef] [PubMed]

30. Ferrari, A.C. Raman spectroscopy of graphene and graphite: Disorder, electron–phonon coupling, doping and nonadiabatic effects. *Solid State Commun.* **2007**, *143*, 47–57. [CrossRef]

31. Gupta, A.; Chen, G.; Joshi, P.; Tadigadapa, S.; Eklund, P.C. Raman Scattering from High-Frequency Phonons in Supported n-Graphene Layer Films. *Nano Lett.* **2006**, *6*, 2667–2673. [CrossRef] [PubMed]

32. Graf, D.; Molitor, F.; Ensslin, K.; Stampfer, C.; Jungen, A.; Hierold, C.; Wirtz, L. Spatially resolved Raman spectroscopy of single- and few-layer graphene. *Nano Lett.* **2007**, *7*, 238–242. [CrossRef] [PubMed]

33. Mignuzzi, S.; Pollard, A.J.; Bonini, N.; Brennan, B.; Gilmore, I.S.; Pimenta, M.A.; Richards, D.; Roy, D. Effect of disorder on Raman scattering of single-layer MoS_2. *Phys. Rev. B* **2015**, *91*, 195411. [CrossRef]

34. Liu, H.; Iskander, A.; Yakovlev, N.L.; Chi, D. Anomalous SiO_2 layer formed on crystalline MoS_2 films grown on Si by thermal vapor sulfurization of molybdenum at elevated temperatures. *Mater. Lett.* **2015**, *160*, 491–495. [CrossRef]

35. Hussain, S.; Singh, J.; Vikraman, D.; Singh, A.K.; Iqbal, M.Z.; Khan, M.F.; Kumar, P.; Choi, D.C.; Song, W.; An, K.S.; et al. Large-area, continuous and high electrical performances of bilayer to few layers MoS_2 fabricated by RF sputtering via post-deposition annealing method. *Sci. Rep.* **2016**, *6*, 30791. [CrossRef] [PubMed]

36. Hussain, S.; Shehzad, M.A.; Vikraman, D.; Khan, M.F.; Singh, J.; Choi, D.; Seo, Y.; Eom, J.; Lee, W.; Jung, J. Synthesis and characterization of large-area and continuous MoS_2 atomic layers by RF magnetron sputtering. *Nanoscale* **2016**, *8*, 4340–4347. [CrossRef] [PubMed]

37. Li, H.; Zhang, Q.; Yap, C.; Tay, B.; Edwin, T.; Olivier, A.; Baillargeat, D. From Bulk to Monolayer MoS_2: Evolution of Raman Scattering. *Adv. Funct. Mater.* **2012**, *22*, 1385–1390. [CrossRef]

38. Liu, Y.J.; Hao, L.Z.; Gao, W.; Liu, Y.M.; Li, G.X.; Xue, Q.Z.; Guo, W.Y.; Yu, L.Q.; Wu, Z.P.; Liu, X.H.; et al. Growth and humidity-dependent electrical properties of bulk-like MoS_2 thin films on Si. *RSC Adv.* **2015**, *5*, 74329–74335. [CrossRef]

39. Ganatra, R.; Zhang, Q. Few-Layer MoS_2: A Promising Layered Semiconductor. *ACS Nano* **2014**, *8*, 4074–4099. [CrossRef] [PubMed]

40. Taylor, C.A.; Wayne, M.F.; Chiu, W.K.S. Residual stress measurement in thin carbon films by Raman spectroscopy and nanoindentation. *Thin Solid Films* **2003**, *429*, 190–200. [CrossRef]

41. Piazza, A.; Giannazzo, F.; Buscarino, G.; Fisichella, G.; Magna, A.; Roccaforte, F.; Cannas, M.; Gelardi, F.M.; Agnello, S. In-situ monitoring by Raman spectroscopy of the thermal doping of graphene and MoS_2 in O_2-controlled atmosphere. *Beilstein J. Nanotech.* **2017**, *8*, 418–424. [CrossRef] [PubMed]

42. Richter, H.; Wang, Z.P.; Ley, L. The one phonon Raman spectrum in microcrystalline silicon. *Solid State Commun.* **1981**, *39*, 625–629. [CrossRef]

43. Ferrari, A.C.; Meyer, J.C.; Scardaci, V.; Casiraghi, C.; Lazzeri, M.; Mauri, F.; Piscanec, S.; Jiang, D.; Novoselov, K.S.; Roth, S. Raman spectrum of graphene and graphene layers. *Phys. Rev. Lett.* **2006**, *97*, 187401. [CrossRef] [PubMed]

44. Zhang, Y.; Zhang, L.; Zhou, C. Review of chemical vapor deposition of graphene and related applications. *Acc. Chem. Res.* **2013**, *46*, 2329–2339. [CrossRef] [PubMed]

45. Chakraborty, B.; Matte, H.; Sood, A.; Rao, C. Layer-dependent resonant Raman scattering of a few layer MoS_2. *J. Raman Spectrosc.* **2013**, *44*, 92–96. [CrossRef]

46. Mercado, E.; Goodyear, A.; Moffat, J.; Cooke, M.; Sundaram, R. A Raman metrology approach to quality control of 2D MoS_2 film fabrication. *J. Phys. D Appl. Phys.* **2017**, *50*, 184005. [CrossRef]

47. Laskar, M.R.; Nath, D.N.; Ma, L.; Lee, E.W. p-type doping of MoS_2 thin films using Nb. *Appl. Phys. Lett.* **2014**, *104*, 092104. [CrossRef]

48. Mcintyre, N.S.; Spevack, P.A.; Beamson, G.; Briggs, D. Effects of argon ion bombardment on basal plane and polycrystalline MoS_2. *Surf. Sci.* **1990**, *237*, L390–L397. [CrossRef]

49. Wagner, C.D.; Riggs, W.M.; Davis, L.E.; Moulder, J.F.; Muilenberg, G.E. *Handbook of X-ray Photoelectron Spectroscopy*; Perkin-Elmer Corporation, Physical Electronics Division Press: Eden Prairie, MN, USA, 1979; p. 190.

50. Al-Shihry, S.S.; Halawy, S.A. Unsupported MoO_3 Fe_2O_3 catalysts: Characterization and activity during 2-propanol decomposition. *J. Mol. Catal. A Chem.* **1996**, *113*, 479–487. [CrossRef]

51. Davis, S.M.; Carver, J.C. Oxygen chemisorption at defect sites in MoS_2 and ReS_2 basal plane surfaces. *Appl. Surf. Sci.* **1984**, *20*, 193–198. [CrossRef]
52. Ahn, C.; Lee, J.; Kim, H.U.; Bark, H.; Jeon, M.; Ryu, G.H.; Lee, Z.; Yeom, G.Y.; Kim, K.; Jung, J.; et al. Low-Temperature Synthesis of Large-Scale Molybdenum Disulfide Thin Films Directly on a Plastic Substrate Using Plasma-Enhanced Chemical Vapor Deposition. *Adv. Mater.* **2015**, *27*, 5223–5229. [CrossRef] [PubMed]
53. Choi, J.G.; Thompson, L.T. XPS study of as-prepared and reduced molybdenum oxides. *Appl. Surf. Sci.* **1996**, *93*, 143–149. [CrossRef]

 *nanomaterials*

Article

α-MoO$_3$ Crystals with a Multilayer Stack Structure Obtained by Annealing from a Lamellar MoS$_2$/g-C$_3$N$_4$ Nanohybrid

Pablo Martín-Ramos [1,*] , Ignacio A. Fernández-Coppel [2] , Manuel Avella [3] and Jesús Martín-Gil [4]

[1] Department of Agricultural and Environmental Sciences, EPS, Instituto de Investigación en Ciencias Ambientales (IUCA), University of Zaragoza, Carretera de Cuarte s/n, 22071 Huesca, Spain

[2] Engineering of Manufacturing Processes group, School of Industrial Engineering, University of Valladolid, C/ Francisco Mendizábal 1, 47014 Valladolid, Spain; ignacio.alonso.fernandez-coppel@uva.es

[3] Unidad de Microscopía Avanzada, Parque Científico UVa, Universidad de Valladolid, Paseo Belén 11, 47011 Valladolid, Spain; um.parque.cientifico@uva.es

[4] Agriculture and Forestry Engineering Department, ETSIIAA, Universidad de Valladolid, Avenida de Madrid 44, 34004 Palencia, Spain; mgil@iaf.uva.es

* Correspondence: pmr@unizar.es; Tel.: +34-974-292-668

Received: 7 July 2018; Accepted: 20 July 2018; Published: 22 July 2018

Abstract: Transition metal oxides and chalcogenides have recently attracted great attention as the next generation of 2-D materials due to their unique electronic and optical properties. In this study, a new procedure for the obtaining of highly crystalline α-MoO$_3$ is proposed as an alternative to vapor-phase synthesis. In this approach, a first reaction between molybdate, citrate and thiourea allowed to obtain MoS$_2$, which—upon calcination at a temperature of 650 $^\circ$C in the presence of g-C$_3$N$_4$—resulted in MoO$_3$ with a definite plate-like shape. The colorless (or greenish) α-MoO$_3$ nanoplates obtained with this procedure featured a multilayer stack structure, with a side-length of 1–2 µm and a thickness of several nanometers viewed along the [010] direction. The nucleation-growth of the crystal can be explained by a two-dimensional layer-by-layer mechanism favored by g-C$_3$N$_4$ lamellar template.

Keywords: α-MoO$_3$; carbon nitride; g-C$_3$N$_4$; molybdenum trioxide; nanoplates; synthesis

1. Introduction

MoO$_3$ is a versatile compound with well-recognized applications in electronics, photo- and electrocatalysis, electrode materials for batteries and pseudocapacitors, gas sensing, superconductors, lubricants, thermoelectric and electrochromic systems, etc., as discussed in detail in the recent review paper by de Castro, et al. [1].

In particular, stoichiometric and intrinsic MoO$_3$ in its α-phase is an n-type semiconductor with a wide bandgap energy of ca. 3 eV (a range from 2.7 to 3.2 eV has been reported), an electron affinity >6 eV and an ionization energy >9 eV [2,3]. Its high work function has led to extensive applications as an anode interfacial layer in electronics (e.g., in solar cells, light-emitting diodes, 2-D field-effect transistors and photodetectors) [2–5].

Orthorhombic α-MoO$_3$ features a layered crystal structure, which offers the possibility to create 2-D morphologies. Those layers are made of atomically thin sheets featuring a thickness of $\approx$0.7 nm, composed of double layers of linked and distorted MoO$_6$ octahedra. In the vertical [010] direction, the distorted MoO$_6$ octahedra are held together by van der Waals' forces, resulting in stratification, while the internal interactions in the octahedra are dominated by strong covalent and ionic bonds [6,7].

Sheet-like orthorhombic α-MoO$_3$ nanostructures are usually prepared by a simple hydrothermal method using ammonium heptamolybdate tetrahydrate and nitric acid [8,9], although both liquid-

and vapor-phase-based alternative approaches have been devised for synthesizing and depositing this oxide. Actually, sputtering is now the most commonly used technique for industrial scale deposition of well-defined, large-area crystalline films of molybdenum oxide [10,11].

Aforementioned approaches have some limitations: physical vapor deposition (PVD) and chemical vapor deposition (CVD) methods have substantial energy requirements, rely on complex equipment and need expert operation skills; on the other hand, most liquid-phase synthesis techniques have problems in terms of relative scalability and repeatability, as well as in terms of their ability to produce molybdenum oxides with high crystallinity, controlled stoichiometry, and morphology.

As most applications require clean and large-sized flakes, this pinpoints a clear need to keep exploring new ways to prepare high quality single-layer transition metal oxides and chalcogenides with high yield. In this work we describe a new procedure to obtain highly ordered multi-layer stacks of molybdenum trioxide (α-MoO$_3$) from a molybdate, citrate and thiourea mixture in propylene carbonate solution upon heating at 650 °C, using carbon nitride (g-C$_3$N$_4$) as a lamellar template. The resulting material may find application, for instance, in the field of clean energy (provided that 2-D α-MoO$_3$ nanosheets have recently been reported to be strong candidates for electrocatalytic hydrogen evolution reaction [12]) or in ultrasensitive plasmonic biosensing [13].

2. Materials and Methods

2.1. Reagents and Synthesis

Ammonium heptamolybdate tetrahydrate ((NH_4)$_6$Mo$_7$O$_{24}$·4H$_2$O, CAS No. 12054-85-2, puriss, $\geq$99%), citric acid monohydrate (C$_6$H$_8$O$_7$·H$_2$O, CAS No. 5949-29-1, ACS reagent, $\geq$99.0%), thiourea (CH$_4$N$_2$S, CAS No. 62-56-6, ACS reagent, $\geq$99.0%) and propylene carbonate (C$_4$H$_6$O$_3$, CAS No. 108-32-7, anhydrous, 99.7%) were purchased from Sigma-Aldrich Química SL (Madrid, Spain), and were used without further purification. g-C$_3$N$_4$ was prepared according to the procedure reported in [14].

2.1.1. Synthesis of MoS$_2$

Firstly, ammonium heptamolybdate tetrahydrate ((NH_4)$_6$Mo$_7$O$_{24}$·4H$_2$O) (4 mmol) was dissolved in 100 mL of distilled water under continuous stirring, and 2 g of citric acid monohydrate (C$_6$H$_8$O$_7$·H$_2$O) were then added to the solution, resulting in a pH of 4. Subsequently, 10 mmol of thiourea were added to the solution mixture and the dispersion was sonicated for 60 min (with a probe-type UIP1000hdT ultrasonicator; Hielscher, Teltow, Germany; 1000 W, 20 kHz) in 10 periods of 2 min each, keeping the temperature below 40 °C. The initial greenish black color changed to dark red and, after heating at 90 °C for 1 h with stirring on a heating magnetic stirrer, color changed from red to black. After centrifugation at 4000 rpm for 1 h, a precipitate was formed, which was washed 3 times with distilled water and ethanol and then dried at 60 °C for 24 h. 1 g of this precipitate was introduced into a 40 mL Teflon-lined stainless steel autoclave and 30 mL of propylene carbonate were added, followed by stirring and heating at 200 °C for 24 h, yielding a solution that would contain [Mo(CH$_4$N$_2$S)$_2$O$_3$]) (or, secondarily, some [(MoO$_2$)$_2$O(C$_6$H$_8$O$_7$)$_2$] as a transient species). By centrifugation of this solution, a precipitate was obtained, in which MoS$_2$ would be the main component [15,16]. This precipitate was dried at 150 °C for 24 h.

Two proposed reaction mechanisms would be:

$$2[Mo(CH_4N_2S)_2O_3] + 8\,O_2 \rightarrow Mo_2S + 3\,SO_2 \uparrow + 4\,CO_2 \uparrow + 8\,H_2O \uparrow + 2\,N_2 \uparrow \qquad (1)$$

$$2[(MoO_2)_2O(C_6H_8O_7)_2] + 2\,CH_4N_2S + 17\,O_2 \rightarrow 2\,Mo_2S + 26\,CO_2 \uparrow + 20\,H_2O \uparrow + 2\,N_2 \uparrow \qquad (2)$$

2.1.2. Synthesis of MoS$_2$/g-C$_3$N$_4$

g-C$_3$N$_4$ was added to MoS$_2$ (1:1 *w/w*, 300 mg of each), the mixture was dispersed in 30 mL of propylene carbonate and stirred at 40 °C for 30 min, followed by sonication for 30 min in four periods of 5 min each, without exceeding 40 °C. By centrifugation of this solution, a precipitate was obtained, which was dried at 150 °C for 24 h to obtain a $MoS_2/g\text{-}C_3N_4$ composite material similar to those previously reported in the literature [17–19].

2.1.3. Synthesis of MoO_3

Nanostructured α-MoO_3 was obtained by heating 500 mg of the hydrothermally synthesized $MoS_2/g\text{-}C_3N_4$ composite in air at 650 °C for 30 min in a Al_2O_3 ceramic crucible with lid in a GVA 12/900 oven (Carbolite Gero, Hope Valley, UK; power: 5.460 kW; heating length: 900 mm; T_{max}: 1200 °C). Thermal heating of the composite at 650 °C gave molybdenum trioxide crystals with a multilayer stacked structure. Carbon nitride oxide, $(g\text{-}C_3N_4)O$, formed from $g\text{-}C_3N_4$, was released as gaseous vapor [20].

$$2\,Mo_2S/g\text{-}C_3N_4 + 9\,O_2 \rightarrow 4\,MoO_3 + 2\,SO_2 \uparrow + 2\,(g\text{-}C_3N_4)O \uparrow \tag{3}$$

2.2. Characterization

The vibrational spectrum in the 400–4000 cm^{-1} spectral range was characterized using a Thermo Scientific (Waltham, MA, USA) Nicolet iS50 Fourier-transform infrared (FT-IR) spectrometer, equipped with an in-built diamond attenuated total reflection (ATR) system, with a 1 cm^{-1} spectral resolution and 64 scans.

The X-ray powder diffraction pattern was obtained with a Bruker (Billerica, MA, USA) D8 Advance powder diffractometer in a Bragg-Brentano geometry, using a silicon crystal low background specimen holder. Data was collected in the $2\theta = 5°$–$80°$ range, with increments of $0.01°$ and an acquisition time per step of 0.5 s.

Scanning electron microscopy (SEM) analysis was carried out with a Tescan (Brno, Czech Republic) Vega3 microscope with BSE (annular, YAG crystal, 0.1 atomic resolution) and SE (Everhart-Thornley type, YAG crystal) detectors, and equipped with a Bruker Quantax 100 Easy energy-dispersive X-ray analysis (EDX) system based on a Bruker Xflash 410 M Silicon Drift Detector, with a 133 eV energy resolution (Mn Ka) @ 100 kcps. Transmission electron (TEM) micrographs and the selected area electron diffraction (SAED) pattern were obtained in a JEM FS2200 HRP microscope (JEOL, Akishima, Tokyo, Japan) operating at 200 kV.

The X-ray photoelectron spectroscopy (XPS) spectrum was collected using a Kratos AXIS UltraDLD instrument (Kratos Analytical Ltd., Manchester, UK) with a monochromatic Al Kα X-ray source (1486.6 eV). For energy calibration, XPS binding energies were referenced to the C 1s peak at 284.6 eV.

The diffuse optical reflectance spectrum (UV-Vis DR) was obtained in a CARY 500 spectrometer (Agilent, Santa Clara, CA, USA) equipped with an integration sphere. The spectrum was recorded in diffuse reflectance mode and transformed by the instrument software to equivalent absorption Kubelka-Munk (K-M) units. The K-M function was plotted as a function of energy and the bandgap value was calculated through the inflection point of this curve.

3. Results and Discussion

3.1. Vibrational Characterization

The reactions between molybdate, citrate and thiourea were tracked by ATR-FTIR spectroscopy. As thiourea was added to molybdate-citrate under ambient conditions, a shift towards lower wavenumbers of the C=S stretching peak (from 731 to 728 cm^{-1}) and of the C–N stretching peak (from 1473 to 1461 cm^{-1}) were observed. These shifts pointed at bond formation between Mo of molybdate and S of thiourea to yield either a molybdenum-thiourea complex or a MoS_2 chalcogenide. However, changes in shape and position of NH_2 and C=O stretching peaks denoted a strong interaction between citrate and the molybdenum-thiourea complex (NH_2 stretching peaks at 2800 and 3700 cm^{-1} were different from those of molybdenum-thiourea and the C=O stretching peak was shifted from 1624 to 1604 cm^{-1}) (Figure 1, dotted line). Presence of some MoS$_2$, even before treatment in the Parr reactor, could be observed in the peak at 482 cm^{-1}, which corresponded to γ_{as} (Mo–S) [21].

Upon addition of g-C$_3$N$_4$ (Figure 1, dashed line), the spectra showed a band at 1204 cm^{-1} due to C/N networks. The peak at 806 cm^{-1} could be either assigned to heptazine ring, to a bending mode of tris-*s*-triazine or to the Mo$_2$–O stretching modes of MoO$_3$. The peak at 541 cm^{-1} was due to νCS vibration. Mo–S characteristic vibration was shifted to 475 cm^{-1}.

Finally, upon treatment at 650 °C, MoS$_2$ was oxidized to MoO$_3$ (Figure 1, solid line). The Mo=O vibration was observed at 1126 cm^{-1}. The peak at 977 cm^{-1} corresponded with the Mo–O bonds. The peak at 815 cm^{-1} was due to the doubly connected bridge-oxygen Mo$_2$–O stretching modes of doubly coordinated oxygen, caused by corner-shared oxygen atoms in common to two MoO$_6$ octahedra [22]. The peak at 556 cm^{-1} was characteristic of stretching vibrations of Mo–O. Some remains of MoS$_2$ were identified by the Mo–S vibration at 471 cm^{-1}.

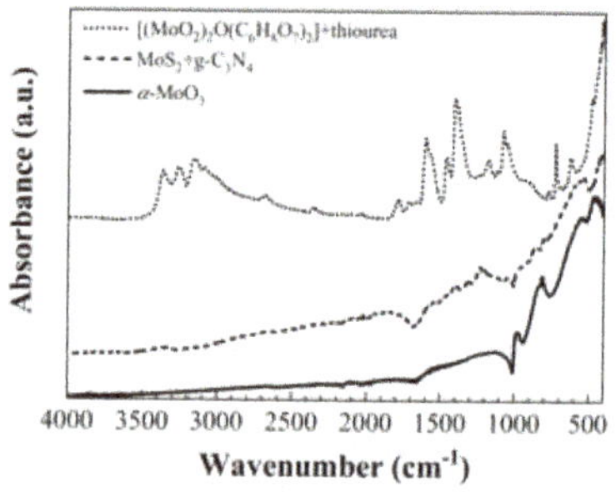

Figure 1. Normalized ATR-FTIR spectra of two intermediate steps of the synthesis and the final α-MoO$_3$ product. An offset has been added for clarity purposes.

3.2. X-Ray Powder Diffraction and Energy-Dispersive X-Ray Spectroscopy Analyses

The X-ray powder diffractogram of the end product for a treatment temperature of 650 °C (Figure 2) matched well the one reported in ICDD crystallographic database for orthorhombic α-MoO$_3$ (PDF 00-005-0508). The positions of the experimental peaks were in good agreement with the simulated diffractogram, albeit with changes in the intensity, which may be explained by preferential orientation resulting from the Bragg-Brentano geometry used in the data collection.

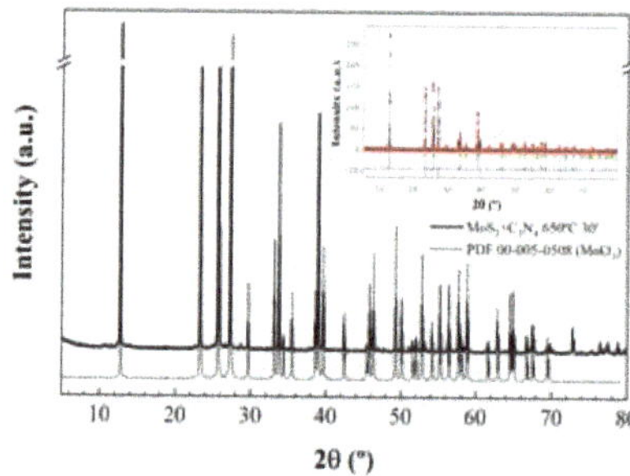

Figure 2. X-ray powder diffraction patterns for the end product upon treatment at 650 °C (solid line) and for orthorhombic α-MoO$_3$ (dotted line). Inset: Rietveld refinement results, using FullProf [23].

The EDX analysis (Figure 3 and Table 1) resulted in a molybdenum to oxygen atomic ratio $A_{Mo:O}$ = 0.29, in reasonably good agreement with the theoretical 0.33 ratio. It also pointed at the presence of aluminum impurities, tentatively ascribed to contamination from the crucible.

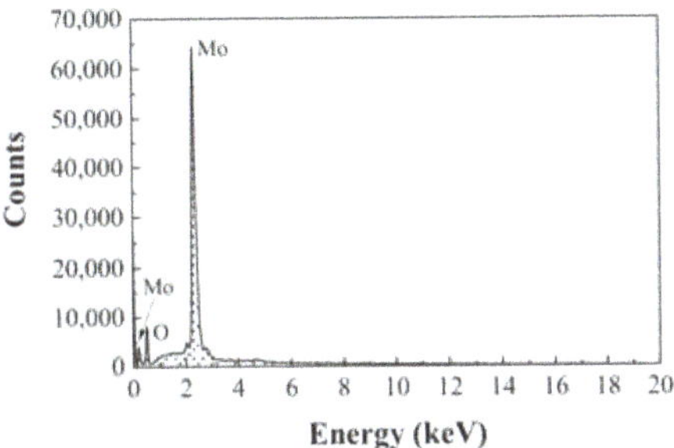

Figure 3. EDX analysis of the end product.

Table 1. Estimated chemical composition of the end product. Data were obtained with EDX in semi-quantitative mode. The errors were automatically calculated by the analysis software.

Element	Series	[wt.%]	[norm. wt.%]	[norm. at.%]	Error in wt.% (3σ)
Oxygen	K-series	25.473	36.624	77.403	9.409
Aluminum	K-series	0.070	0.100	0.126	0.089
Sulfur *	K-series	0.172	0.247	0.124	0.097
Molybdenum	L-series	43.839	63.029	22.210	4.732

* The percentage assigned to sulfur may be ascribed to limitations of the software in the discrimination of molybdenum and sulfur by peak deconvolution.

3.3. Scanning and Transmission Electron Microscopy Analyses

Molybdenum trioxide obtained by the procedure reported above was a greenish-white crystalline material. SEM micrographs revealed a multilayer stack structure built from planar crystals, either 1×1 μm or 2×2 μm in size (Figure 4a–c). The shape of the crystals was similar to those reported by Wang, et al. [24], [25] or Vila, et al. [26], corresponding to α-MoO$_3$, and was very different from that obtained by calcination of commercial molybdic acid, MoO$_3 \cdot$H$_2$O (Figure 4d).

Figure 4. (**a–c**) SEM micrographs of α-MoO$_3$ crystals with a multi-layer stack structure at different magnifications (20,000×, 50,000× and 120,000×, respectively); (**d**) SEM image of MoO$_3$ crystals obtained by calcination of commercial molybdic acid (MoO$_3 \cdot$H$_2$O); (**e**) SEM micrograph of α-MoO$_3$ stacking with 44 layers.

In the SEM micrographs presented above, the number of stacked layers varied between 3 and 44. For a stacking of 1 µm of thickness (Figure 4e), 44 layers could be discerned. The spacing between layers derived from the cross-section SEM images (25 nm) was around 18 times the thickness of two double-layers within a unit cell of the orthorhombic α-MoO$_3$ crystal.

Figure 5a,b shows TEM micrographs of the α-MoO$_3$ nanoplates, similar to those obtained, for instance, by calcination of h-MoO$_3$ microrods [27]. The SAED pattern (Figure 5c) was indexed to correspond with the (002), (202), and (200) crystallographic planes, which were specified as orthorhombic α-MoO$_3$ [28], in accordance with the XRD analysis.

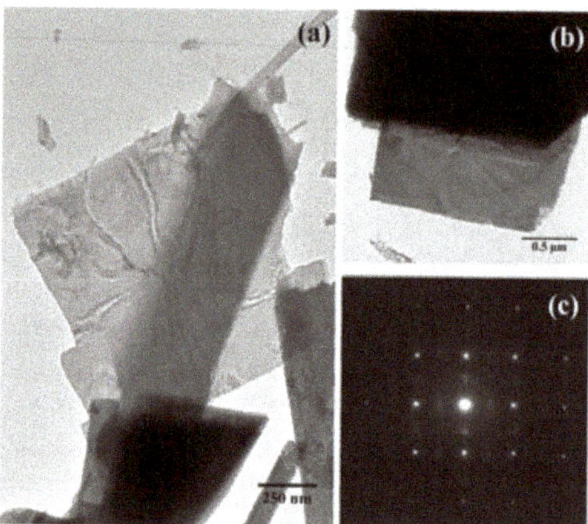

Figure 5. (a,b) TEM micrographs of the α-MoO$_3$ nanoplates; (c) SAED pattern.

3.4. Surface Characterization

Figure 6 shows the Mo (3*d*) XPS spectrum of the α-MoO$_3$ sample. The doublet at 232.88 and 235.99 eV are attributed to the binding energies of the $3d_{5/2}$ and $3d_{3/2}$ electrons of Mo^{6+}, respectively, in good agreement with previous reports of Mo^{6+} state of α-MoO$_3$ [29,30].

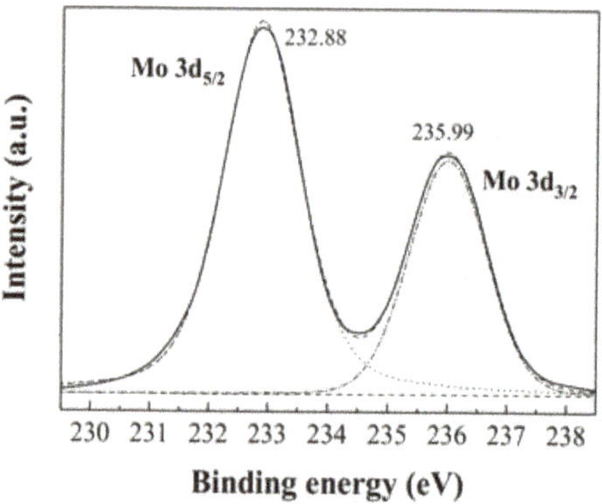

Figure 6. High resolution scan of Mo 3d doublet core levels of α-MoO$_3$.

3.5. Optical Properties

To examine the optical properties of the sample, its UV-Vis DR spectrum was recorded over the 200–800 nm wavelength range at room temperature. As shown in the inset in Figure 7, the optical band gap was found to be ~3.1 eV, which is in good agreement with values reported in the literature [25,31].

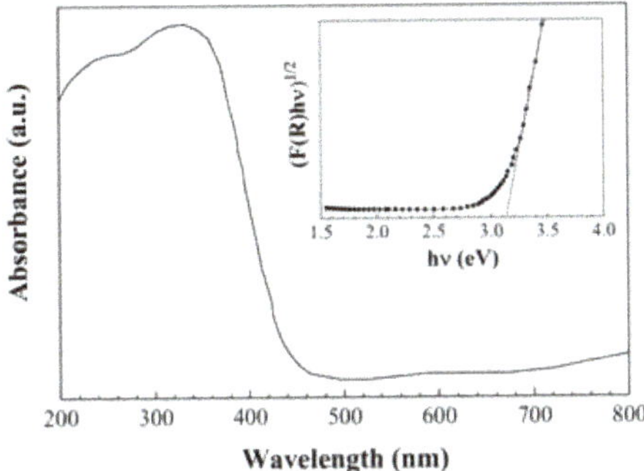

Figure 7. Diffuse reflectance spectrum and band gap energy (inset) of α-MoO$_3$.

4. Conclusions

A novel method for the preparation of high quality α-MoO$_3$ was proposed, based on the use of g-C$_3$N$_4$ as a lamellar template for the calcination of MoS$_2$ (previously obtained from molybdate, citrate and thiourea) at 650 °C. The resulting orthorhombic molybdenum oxide was characterized by X-ray powder diffraction, ATR-FTIR, SEM, TEM, EDX, XPS and UV-Vis DR. X-ray powder diffraction data confirmed the good crystallinity of the obtained product, while the micrographs evinced the presence of well-defined large nanoplates, comparable to those obtained by vapor-phase synthesis techniques. The proposed procedure may thus pose an alternative to PVD and CVD methods, as it can overcome some of their limitations of in terms of energy requirements and equipment, and to conventional liquid-phase synthesis techniques, provided that it can result in higher crystallinity.

Author Contributions: Conceptualization, J.M.-G.; Formal analysis, P.M.-R. and J.M.-G.; Funding acquisition, P.M.-R.; Investigation, P.M.-R., I.A.F.-C., M.A. and J.M.-G.; Methodology, J.M.-G.; Resources, J.M.-G.; Supervision, J.M.-G.; Validation, P.M.-R. and I.A.F.-C.; Visualization, P.M.-R., M.A. and I.A.F.-C.; Writing—original draft, P.M.-R., I.A.F.-C. and J.M.-G.; Writing—review & editing, P.M.-R.

Funding: This research was funded by Santander Universidades through the "Becas Iberoamérica Jóvenes Profesores e Investigadores, España" scholarship program. The APC was funded by IUCA, Universidad de Zaragoza.

Acknowledgments: Access to TAIL-UC facility funded under QREN-Mais Centro project ICT-2009-02-012-1980 is gratefully acknowledged.

Conflicts of Interest: The authors declare no conflict of interest. The funders had no role in the design of the study; in the collection, analyses, or interpretation of data; in the writing of the manuscript, and in the decision to publish the results.

References

1. de Castro, I.A.; Datta, R.S.; Ou, J.Z.; Castellanos-Gomez, A.; Sriram, S.; Daeneke, T.; Kalantar-zadeh, K. Molybdenum oxides—From fundamentals to functionality. *Adv. Mater.* **2017**, *29*. [CrossRef] [PubMed]

2. Meyer, J.; Hamwi, S.; Kröger, M.; Kowalsky, W.; Riedl, T.; Kahn, A. Transition metal oxides for organic electronics: Energetics, device physics and applications. *Adv. Mater.* **2012**, *24*, 5408–5427. [CrossRef] [PubMed]

3. Qu, Q.; Zhang, W.-B.; Huang, K.; Chen, H.-M. Electronic structure, optical properties and band edges of layered MoO$_3$: A first-principles investigation. *Comput. Mater. Sci.* **2017**, *130*, 242–248. [CrossRef]

4. Alsaif, M.M.Y.A.; Chrimes, A.F.; Daeneke, T.; Balendhran, S.; Bellisario, D.O.; Son, Y.; Field, M.R.; Zhang, W.; Nili, H.; Nguyen, E.P.; et al. High-performance field effect transistors using electronic inks of 2D molybdenum oxide nanoflakes. *Adv. Funct. Mater.* **2016**, *26*, 91–100. [CrossRef]

5. Balendhran, S.; Deng, J.; Ou, J.Z.; Walia, S.; Scott, J.; Tang, J.; Wang, K.L.; Field, M.R.; Russo, S.; Zhuiykov, S.; et al. Enhanced charge carrier mobility in two-dimensional high dielectric molybdenum oxide. *Adv. Mater.* **2013**, *25*, 109–114. [CrossRef] [PubMed]

6. Kalantar-zadeh, K.; Tang, J.; Wang, M.; Wang, K.L.; Shailos, A.; Galatsis, K.; Kojima, R.; Strong, V.; Lech, A.; Wlodarski, W.; et al. Synthesis of nanometre-thick MoO₃ sheets. *Nanoscale* **2010**, *2*, 429–433. [CrossRef] [PubMed]

7. Ji, F.; Ren, X.; Zheng, X.; Liu, Y.; Pang, L.; Jiang, J.; Liu, S. 2D-MoO₃ nanosheets for superior gas sensors. *Nanoscale* **2016**, *8*, 8696–8703. [CrossRef] [PubMed]

8. Rathnasamy, R.; Thangamuthu, R.; Alagan, V. Sheet-like orthorhombic MoO₃ nanostructures prepared via hydrothermal approach for visible-light-driven photocatalytic application. *Res. Chem. Intermed.* **2017**, *44*, 1647–1660. [CrossRef]

9. Shahab ud, D.; Ahmad, M.Z.; Qureshi, K.; Bhatti, I.A.; Zahid, M.; Nisar, J.; Iqbal, M.; Abbas, M. Hydrothermal synthesis of molybdenum trioxide, characterization and photocatalytic activity. *Mater. Res. Bull.* **2018**, *100*, 120–130. [CrossRef]

10. Yao, D.D.; Ou, J.Z.; Latham, K.; Zhuiykov, S.; O'Mullane, A.P.; Kalantar-zadeh, K. Electrodeposited α-MoO₃- and β-Phase MoO₃ Films and Investigation of Their Gasochromic Properties. *Cryst. Growth Des.* **2012**, *12*, 1865–1870. [CrossRef]

11. Chang, W.-C.; Qi, X.; Kuo, J.-C.; Lee, S.-C.; Ng, S.-K.; Chen, D. Post-deposition annealing control of phase and texture for the sputtered MoO₃ films. *CrystEngComm* **2011**, *13*, 5125–5132. [CrossRef]

12. Datta, R.S.; Haque, F.; Mohiuddin, M.; Carey, B.J.; Syed, N.; Zavabeti, A.; Zhang, B.; Khan, H.; Berean, K.J.; Ou, J.Z.; et al. Highly active two dimensional α-MoO₃₋ₓ for the electrocatalytic hydrogen evolution reaction. *J. Mater. Chem. A* **2017**, *5*, 24223–24231. [CrossRef]

13. Zhang, B.Y.; Zavabeti, A.; Chrimes, A.F.; Haque, F.; O'Dell, L.A.; Khan, H.; Syed, N.; Datta, R.; Wang, Y.; Chesman, A.S.R.; et al. Degenerately hydrogen doped molybdenum oxide nanodisks for ultrasensitive plasmonic biosensing. *Adv. Funct. Mater.* **2018**, *28*, 1706006. [CrossRef]

14. Dante, R.C.; Martín-Ramos, P.; Correa-Guimaraes, A.; Martín-Gil, J. Synthesis of graphitic carbon nitride by reaction of melamine and uric acid. *Mater. Chem. Phys.* **2011**, *130*, 1094–1102. [CrossRef]

15. Vikraman, D.; Akbar, K.; Hussain, S.; Yoo, G.; Jang, J.-Y.; Chun, S.-H.; Jung, J.; Park, H.J. Direct synthesis of thickness-tunable MoS₂ quantum dot thin layers: Optical, structural and electrical properties and their application to hydrogen evolution. *Nano Energy* **2017**, *35*, 101–114. [CrossRef]

16. Vattikuti, S.V.P.; Byon, C. Synthesis and Characterization of Molybdenum Disulfide Nanoflowers and Nanosheets: Nanotribology. *J. Nanomater.* **2015**, *2015*, 1–11. [CrossRef]

17. Wang, J.; Guan, Z.; Huang, J.; Li, Q.; Yang, J. Enhanced photocatalytic mechanism for the hybrid g-C₃N₄/MoS₂ nanocomposite. *J. Mater. Chem. A* **2014**, *2*, 7960–7966. [CrossRef]

18. Li, J.; Liu, E.; Ma, Y.; Hu, X.; Wan, J.; Sun, L.; Fan, J. Synthesis of MoS₂/g-C₃N₄ nanosheets as 2D heterojunction photocatalysts with enhanced visible light activity. *Appl. Surf. Sci.* **2016**, *364*, 694–702. [CrossRef]

19. Ge, L.; Han, C.; Xiao, X.; Guo, L. Synthesis and characterization of composite visible light active photocatalysts MoS₂–g-C₃N₄ with enhanced hydrogen evolution activity. *Int. J. Hydrogen Energy* **2013**, *38*, 6960–6969. [CrossRef]

20. Kharlamov, A.; Bondarenko, M.; Kharlamova, G.; Gubareni, N. Features of the synthesis of carbon nitride oxide (g-C₃N₄)O at urea pyrolysis. *Diamond Relat. Mater.* **2016**, *66*, 16–22. [CrossRef]

21. Nagaraju, G.; Tharamani, C.N.; Chandrappa, G.T.; Livage, J. Hydrothermal synthesis of amorphous MoS₂ nanofiber bundles via acidification of ammonium heptamolybdate tetrahydrate. *Nanoscale Res. Lett.* **2007**, *2*, 461–468. [CrossRef] [PubMed]

22. Siciliano, T.; Tepore, A.; Filippo, E.; Micocci, G.; Tepore, M. Characteristics of molybdenum trioxide nanobelts prepared by thermal evaporation technique. *Mater. Chem. Phys.* **2009**, *114*, 687–691. [CrossRef]

23. Rodríguez-Carvajal, J. Recent developments of the program FULLPROF. *Comm. Powder Diffr. (IUCr) Newslett.* **2001**, *26*, 12–19.

24. Wang, T.; Li, J.; Zhao, G. Synthesis of MoS₂ and MoO₃ hierarchical nanostructures using a single-source molecular precursor. *Powder Technol.* **2014**, *253*, 347–351. [CrossRef]

25. Lou, S.N.; Yap, N.; Scott, J.; Amal, R.; Ng, Y.H. Influence of MoO₃ (110) crystalline plane on its self-charging photoelectrochemical properties. *Sci. Rep.* **2014**, *4*, 7428. [CrossRef] [PubMed]

26. Vila, M.; Díaz-Guerra, C.; Jerez, D.; Lorenz, K.; Piqueras, J.; Alves, E. Intense luminescence emission from rare-earth-doped MoO₃ nanoplates and lamellar crystals for optoelectronic applications. *J. Phys. D Appl. Phys.* **2014**, *47*, 35. [CrossRef]

27. Wongkrua, P.; Thongtem, T.; Thongtem, S. Synthesis of h- and α-MoO$_3$ by refluxing and calcination combination: Phase and morphology transformation, photocatalysis, and photosensitization. *J. Nanomater.* **2013**, *2013*, 1–8. [CrossRef]

28. Klinbumrung, A.; Thongtem, T.; Thongtem, S. Characterization of orthorhombic α-MoO$_3$ microplates produced by a microwave plasma process. *J. Nanomater.* **2012**, *2012*, 1–5. [CrossRef]

29. Xia, T.; Li, Q.; Liu, X.; Meng, J.; Cao, X. Morphology-controllable synthesis and characterization of single-crystal molybdenum trioxide. *J. Phys. Chem. B* **2006**, *110*, 2006–2012. [CrossRef] [PubMed]

30. Patel, S.K.S.; Dewangan, K.; Gajbhiye, N.S. Synthesis and room temperature d$_0$ ferromagnetic properties of α-MoO$_3$ nanofibers. *J. Mater. Sci. Technol.* **2015**, *31*, 453–457. [CrossRef]

31. Chithambararaj, A.; Bose, A.C. Hydrothermal synthesis of hexagonal and orthorhombic MoO$_3$ nanoparticles. *J. Alloys Compd.* **2011**, *509*, 8105–8110. [CrossRef]

MDPI

St. Alban-Anlage 66

4052 Basel

Switzerland

Tel. +41 61 683 77 34

Fax +41 61 302 89 18

www.mdpi.com

Nanomaterials Editorial Office

E-mail: nanomaterials@mdpi.com

www.mdpi.com/journal/nanomaterials